Results from the First Mathematics Assessment of the National Assessment of Educational Progress

THOMAS CARPENTER
TERRENCE G. COBURN
ROBERT E. REYS
JAMES W. WILSON

with the assistance of
MARY K. CORBITT

This publication is a report from the NCTM Project for Interpretive Reports on National Assessment. Chapters I and II were prepared by the NAEP staff.

NATIONAL COUNCIL OF TEACHERS OF MATHEMATICS

Library of Congress Cataloguing in Publication Data:

National Council of Teachers of Mathematics.
    Results from the first mathematics assessment
of the National Assessment of Educational Progress.

    "A report from the NCTM Project for Interpretive
Reports on National Assessment."
    Bibliography: p.
    1. Mathematics—Study and teaching—United
States. 2. Mathematical ability—Testing.
3. National Assessment of Educational Progress
(Project) I. Carpenter, Thomas P. II. National
Assessment of Educational Progress (Project).
III. Title.
QA13.N36   1978   510'.7'1073   78-2345
ISBN 0-87353-123-X

    The Project presented or reported herein was financed in whole or in
part under a contract with the National Assessment of Educational
Progress, a project of the Education Commission of the States, using
funds from the Office of the Assistant Secretary for Education, Department of Health, Education, and Welfare. However, the opinions expressed herein do not necessarily reflect the position or policy of the
Office of the Assistant Secretary for Education, the Education Commission of the States, or the National Assessment of Educational Progress.
No official endorsement by the Office of Education, the Education Commission of the States, or the National Assessment of Educational Progress should be inferred.

Printed in the United States of America

Table of Contents

Prologue

This monograph is a comprehensive presentation and discussion of the 1972-73 National Assessment mathematics data. It is a report from the NCTM Project for Interpretive Reports on National Assessment. Previous NCTM Project reports have appeared as journal articles. This monograph presents an overview of the whole mathematics assessment, examining data from all four age groups and for all content areas. The previous journal article reports have, of necessity, been on selected exercises, data and age groups.

In 1971, NCTM President, H. Vernon Price appointed an Ad Hoc Committee to Interpret Findings of NAEP. James Wilson was chairman of the committee; Jack Price and Vernon Hood were the other members. The charge was to work with National Assessment staff and to prepare a report for mathematics teachers, administrators, and interested lay persons. The committee met with National Assessment Staff and over a period of discussion and negotiation agreed to assist National Assessment with the analysis of mathematics data and to organize a team of mathematics educators to prepare interpretive reports. The National Assessment mandate was to present its data in NAEP reports and leave interpretation to professional groups such as NCTM. This mandate has changed now, but the 1972-73 mathematics assessment was analyzed and reported cooperatively.

The NCTM Project for Interpretive Reports used a team of four people: James Wilson, Robert Reys, Terry Coburn, and Thomas Carpenter. The first draft of this monograph was prepared during the summer 1974. This draft was utilized then to prepare summary manuscripts for the _Arithmetic Teacher_ and the _Mathematics Teacher_. These journal articles were published in 1975-76. The monograph was revised and edited by Mary K. Corbitt.

We feel the information contained in the monograph is of some currency even though there is a long time lag between the assessment and the report. First, there are considerable data presented in this monograph that were not included in the previous summaries. Second, the _second_ mathematics assessment is in the field-stage during 1977-78. This monograph will provide a comprehensive account of the baseline data from the first assessment.

Thus we present this monograph on the 1972-73 mathematics assessment as a reminder of where we were five years ago in mathematics and in anticipation of the assessment now underway.

I

Overview of National Assessment

History and Purpose

When the United States Office of Education was founded in 1867, one charge set before its commissioner was to determine the nation's progress in education. That century-old charge is only now being answered by the National Assessment of Educational Progress, a project of the National Center for Education Statistics under contract to the Education Commission of the States.

By the early 1960's, the average annual expenditure of public funds for the formal education of young Americans was $30 billion.[1] Yet criticisms of the educational system abounded. Defenders of the educational establishment found it increasingly difficult to provide evidence that the schools were satisfactorily meeting the educational needs of a modern, technological society. The only readily available measures of educational quality resulting from this public investment of funds were input measures such as teacher-student ratios and per-pupil expenditures. The tenuous assumption was made that the quality of educational outcomes-- what students do or do not know and can or cannot do--was directly related to the quality of the inputs to the educational system. There has been no conclusive empirical evidence to support this assumption. The typical standardized achievement tests administered by schools or states provided scores whereby one student could be compared with other students. Such information was useful in categorizing students; however, it provided little information about what students were or were not actually learning. There had been no direct assessment of educational outcomes.

This insufficiency of information became the concern of Francis Keppel, United States Commissioner of Education (1962-65). He initiated a series of conferences to explore ways to provide the necessary information. In 1964, as a result of these conferences, John W. Gardner, president of the Carnegie Corporation, asked a distinguished group of educators and other concerned persons to form the Exploratory Committee on Assessing the Progress of Education (ECAPE). This committee, chaired by Dr. Ralph W. Tyler, who had been involved since the earliest conferences, was to examine the possibility of conducting an assessment of educational attainments on a national basis.

After much study, ECAPE decided that it was feasible to inaugurate an assessment project to fill the information gap regarding the quality

---

[1]Based on United States Census Bureau data. Actual amounts expended were as follows: 1960, $24.7 billion; 1962, $29.4 billion; 1964, $35.9 billion.

of educational outcomes by a periodic assessment of the knowledge, skill, understanding and attitudes in ten learning areas[2] at four age levels (9, 13, 17 and adults--ages 26-36). This project, named the Committee on Assessing the Progress of Education (CAPE), began its charge under the auspices of the Carnegie Corporation by assessing the learning areas of Science, Citizenship and Writing in the spring of 1969. Later that same year, the project came under the auspices of the Education Commission of the States. Funding and monitoring were transferred to the Education Division of the U.S. Department of Health, Education, and Welfare, and the project was renamed the National Assessment of Educational Progress (NAEP).

Goals of the National Assessment

National Assessment provides information to educational decision-makers and practitioners that can be used to identify educational problem areas, to establish educational priorities and to determine the national progress in education. To do so, the Assessment must remain flexible enough to accommodate possible extensions, refinements and modifications. The following goals have been established for the project by the National Assessment Policy Committee, the Analysis Advisory Committee and the Assessment staff.

Goal I: To measure change in the educational attainments of young Americans.

Goal II: To make available on a continuing basis comprehensive data on the educational attainments of young Americans.

Goal III: To utilize the capabilities of National Assessment to conduct special interest "probes" into selected areas of educational attainment.

Goal IV: To provide data, analyses and reports understandable to, interpretable by, and responsive to the needs of a variety of audiences.

Goal V: To encourage and facilitate interpretive studies of NAEP data, thereby generating implications useful to educational practitioners and decision-makers.

Goal VI: To facilitate the use of NAEP technology at state and local levels when appropriate.

Goal VII: To continue to develop, test and refine the technologies necessary for gathering and analyzing NAEP achievement data.

---

[2]Art, Career and Occupational Development, Citizenship, Literature, Mathematics, Music, Reading, Science, Social Studies and Writing.

Goal VIII: To conduct an ongoing program of research and operational
studies necessary for the resolution of problems and
refinement of the NAEP model. (Implicit in this goal
is the conduct of research to support previously mentioned
goals.)

Methodology

To measure the nation's educational progress, National Assessment
estimates the percentage of respondents (at any of four age levels) who
are able to acceptably answer a question or perform a task. Each question
or task (called an exercise) reflects an educational goal or objective.
The exercises are administered to scientifically selected samples which
take into account such variables as size and type of community, race and
geographic region. Students are sampled at three age levels that represent educational milestones attained by most students: age 9, when most
students have been exposed to the basic program of primary education;
age 13, when most students have finished their elementary school education;
and age 17, when most students are near completing their secondary education.
To accurately reflect the skills, knowledge, and attitudes of the 17-year-
olds, National Assessment also samples 17-year-olds not enrolled in school.
Adults (ages 26 to 35) are assessed to determine the skills, knowledge
and attitudes of those who have completed their formal education and have
probably been away from school for a number of years. The samples are
designed so that valid inferences can be made about the populations from
which the samples were selected.

National Assessment does not use total performance scores for
individual respondents because its main concern is how various groups
of individuals perform on specific exercises. Thus, it is not necessary
for each respondent to take every exercise. The exercises are sorted
into booklets, and each in-school respondent takes only one booklet.
The various booklets are administered to statistically equivalent samples
so that group comparisons can be made across booklets. This technique
allows National Assessment to assess performance on far more exercises
than would be possible in the usual one-hour testing situation and provides
broader coverage of the assessment objectives for each learning area.

Since individuals are not ranked according to their performance on
the assessment materials, National Assessment does not emphasize the use
of exercises with high discrimination power. The aim of the project is
to describe attainment; this is best accomplished if the exercises used
cover the entire spectrum of difficulty, from very easy tasks to the
most difficult.

While multiple-choice exercises predominate, many open-ended exercises
requiring anywhere from a few words to a long essay as an answer are included
in each assessment. Exercise writers are instructed to use the exercise
format that provides the best and most direct measure of the objectives
being assessed. They are encouraged to develop exercises that employ
the use of pictures, tapes, films or practical, everyday items as stimuli.
In many assessments, individual interviews and observations of the respondents' problem-solving techniques supplement the usual data-collecting

procedures. For example, in Music, respondents were asked to sing a song or perform on an instrument; in Science, respondents were asked to conduct a small experiment; in Social Studies, respondents were asked to interpret an election ballot.

National Assessment regards positive attitudes toward or opinions about the various learning areas as important educational attainments. Therefore, affective exercises and attitude survey questions are also included in most assessments.

Assessment exercises are administered either to individual or to small groups (generally less than 25) by specially trained personnel. Exercises are administered to all out-of-school respondents in a one-to-one situation. Some of the in-school respondents are assessed in a one-to-one situation. This proportion varies according to age level, learning area and assessment year. Exercises specifically designed for individual administration include those having unusual stimuli or requiring something other than a written response. In group administrations, instructions and the exercises themselves are presented to the respondents on paced tape recordings to assure complete and uniform presentation of instruction and to give those who have a reading problem a chance to hear the exercise as they are reading it.

To report the nation's educational progress, the project releases up to one-half of the exercises administered in a learning area. Released exercises allow the public to evaluate the exercises and the accompanying data. The other exercises are kept confidential and are used to assess performance changes over time.

How do NAEP test booklets differ from standardized achievement tests? Standardized achievement tests are norm-referenced; National Assessment tests are content or objective-referenced. On a standardized achievement test, each respondent takes every exercise, receives a score for his performance, and is ranked on the basis of that score with respect to a reference group. No respondent takes all of the exercises National Assessment uses to assess a learning area, no respondent receives a score, and emphasis is placed on the performance of groups of respondents on specific exercises. Standardized achievement test items are usually limited to a multiple-choice format; National Assessment employs a wide variety of exercise formats. Standardized tests usually focus on the cognitive domain; National Assessment usually includes exercises relating to the affective domain as well. A respondent is required to read the items himself when taking most standardized tests; Assessment exercises are read to respondents by a paced tape or the exercises administrator in a interview situation except during the Reading Assessment. The items on a standardized achievement test are rarely, if ever, made public; the Assessment releases up to half of the exercises used in an assessment to accompany the data.

The Assessment Cycle

A single cycle of a learning area assessment, from objectives development or redevelopment to completion of the basic technical reporting of the data, requires approximately four years. Development takes one and one-half years, one and one-half years are spent preparing for and performing data collection, and one year is required for preliminary analysis and basic reporting.

During the first nine years, the following learning areas have been assessed:

    1969   -- Science, Citizenship, Writing
    1970-71-- Reading, Literature
    1971-72-- Music, Social Studies
    1972-73-- Science, Mathematics
    1973-74-- Career and Occupational Development, Writing
    1974-75-- Art, Reading
    1975-76-- Citizenship-Social Studies
    1976-77-- Science, Basic Life Skills
    1977-78-- Mathematics, Consumerism

Future assessments include the regular learning areas as well as small-scale assessments called "special probes." Special probes are topics of special interest which are assessed once and usually at one age (normally 17-year-olds). All the exercises are released if the area is not planned to be reassessed at a later date. Special probes that have been completed by National Assessment are Basic Life Skills and Consumerism.

II

The First Assessment of Mathematics

The National Assessment of Educational Progress conducted the first
Assessment of Mathematics during the 1972-73 school year. The objectives
for this assessment had been developed by two educational testing con-
tractors: Educational Testing Service and Psychological Corporation.
Each contractor independently developed a set of objectives, relying on
its staff, mathematicians and mathematics educators. The final sets
of objectives were reviewed by panels of interested lay citizens to decide
which set to use in the assessment. Panel members were evenly divided
in their preference for the two sets of objectives. In the absence of
a strong preference, the Psychological Corporation was asked to continue
the development of objectives.   In 1968, the Psychological Corporation
completed its revision of the objectives. The revision, together with
objectives selected from the Educational Testing Service's version, was
compiled into a final statement of objectives for the first Assessment
of Mathematics. A booklet containing the statement was published in
1970.

When the objectives for mathematics were first formulated, they
were compared to other statements of objectives that had appeared in
the mathematics education literature during the preceeding twenty-five
years. The objectives for the first assessment were consistent with
objectives appearing in the literature. This outcome was both desired
and expected since one criterion for the National Assessment objectives
was that they be central to prevailing teaching efforts.

A three-dimensional classification scheme was used to categorize
the mathematics objectives for the first assessment. One dimension of
the scheme was Uses of Mathematics, which was divided into three major
categories:

1) Social Mathematics (the mathematics needed for personal living
    and effective citizenship in our society)

2) Technical Mathematics (the mathematics necessary for various
    skilled jobs and professions)

3) Academic Mathematics (the formally structured mathematics that
    provides the basis for an understanding of various mathematical
    processes)

Another dimension of the matrix was Content. The content areas were:

1) Number and Numeration Concepts
2) Properties of Numbers and Operations
3) Arithmetic Computations
4) Sets
5) Estimation and Measurement

9

6) Exponents and Logarithms
7) Algebraic Expressions
8) Equations and Inequalities
9) Functions
10) Probability and Statistics
11) Geometry
12) Trigonometry
13) Mathematical Proof
14) Logic
15) Miscelleneous Topics
16) Business and Consumer Mathematics
17) Attitude and Interest

The third dimension of the classification scheme consisted of six cognitive Objectives or Abilities:

1) To recall and/or recognize definitions, facts and symbols
2) To perform mathematical manipulations
3) To understand mathematical concepts and processes
4) To solve mathematical problems--social, technical, and academic
5) To use mathematics and mathematical reasoning to analyze problem situations, define problems, formulate hypotheses, make decisions, and verify results
6) To appreciate and use mathematics

During the development and review of the exercises, the content and ability dimensions of the classification scheme were the most useful. The exercise developers tended not to use the third dimension, uses of mathematics, when classifying exercises. This third dimension tended to pose too many restrictions on exercise development to make its use worthwhile.

Although the exercises were classified by content and ability, not all content areas or abilities were assessed equally. Certain content topics were purposely measured in more detail than others. Although the objectives were intended to include all the mathematics taught in the nation's schools, it was impossible to measure every objective in depth. Little emphasis, for example, was placed on the topics of trigonometry and logic. The content area of "attitude and interest" and the related ability of "appreciation and use of mathematics" were not measured because the exercises developed to assess these were considered inadequate.

After the exercises had undergone extensive reviews by mathematicians, mathematics educators, and concerned lay persons, the exercises to be used in the assessment were selected and arranged in booklets. The 13-year-olds were assessed during October and November of 1972, 9-year-olds during January and February of 1973, the in-school 17-year-olds during March and April of 1973, and the out-of school 17-year-olds (dropouts and early graduates) and adults were assessed from January through June of 1973. The majority of the exercises were open-ended (free response) and required the respondent to supply the answer. Respondents were asked to do all written calculations or "scratch work" in the assessment booklets. The responses to the open-ended exercises, including any work the respondents did in the booklets were tabulated in various scoring categories. These tabulations revealed the percentages of respondents making particular types of errors and thus provided some diagnostic information about common mathematical errors. In order to reduce guessing, respondents were instructed to write the words, "I don't know" on the answer line or to fill in the oval beside the "I don't know" choice for multiple-choice exercises if they felt that they did not know the answer to a particular exercise.

Most assessment materials were administered to groups of eight to twelve respondents at a time, using a paced audio tape to standardize administrations. However, some exercises were administered in a "one-to-one" interview situation in which the exercises were presented to the respondent by a trained exercise administrator. These individually administered exercises were used to elicit responses that would be difficult to observe in a group situation, such as the process a respondent used to solve a problem. Although the out-of-school respondents were not assessed in group settings, the audio tape was used to standardized administrations.

Approximately half of the exercises used in the Mathematics Assessment were released. These exercises were included in various mathematics reports and made available to groups and individuals for their own uses. The unreleased exercises were used again in the second Assessment of Mathematics to measure changes in educational attainments. A general survey of the results of the first Mathematics Assessment is provided in The First National Assessment of Mathematics: An Overview, Report 04-MA-00.[1] The text of each released exercise and accompanying documentation including results can be found in the Mathematics Technical Report: Exercise Volume, Report 04-MA-20. Data are provided for all of the mathematics exercises, but the exact text and scoring guides are provided for the released exercises only.

Results concerning computational abilities of young Americans are presented and discussed in a special report, Math Fundamentals: Selected Results from the First National Assessment of Mathematics, Report 04-MA-01. The results concerning consumer mathematics are presented and discussed in Consumer Math: Selected Results from the First National Assessment of Mathematics, Report 04-MA-02. In addition, National Assessment has produced computer data tapes containing respondent-level data for the released mathematics exercises. These data tapes are available to any researcher who wishes to conduct further analysis of the data.[2]

---

[1] National Assessment reports can be ordered directly from Public Information Department, National Assessment of Educational Progress, Suite 700, Lincoln Towers, 1860 Lincoln Street, Denver, Colorado 80295, or through the Superintendent of Documents, U. S. Government Printing Office, Washington, D.C. 20402.

[2] Data tapes are available at a charge, through the Data Processing Department, National Assessment of Educational Progress, Suite 700, 1860 Lincoln Street, Denver, Colorado 80295.

National Assessment has worked closely with a panel of mathematics educators from the National Council of Teachers of Mathematics (NCTM) who studied the data in order to draw implications from the results of the first Mathematics Assessment. The NCTM panel presented summaries of its findings in the October 1975 issues of The Arithmetic Teacher[3] and The Mathematics Teacher.[4] Additional brief articles on specific content topics were presented in the October 1975 through May 1976 The Arithmetic Teacher. These articles suggest some of the ways mathematics teachers might use information from the first assessment.

---

[3]Carpenter, Thomas R., Terrence G. Coburn, Robert E. Reys, James W. Wilson, "Results and Implications of the NAEP Mathematics Assessment: Elementary School," The Arithmetic Teacher, Vol. 22, No. 6 (October 1975), pp. 438-450.

[4]Carpenter, T. P., Terrence G. Coburn, Robert E. Reys, James W. Wilson, "Results and Implications of the NAEP Mathematics Assessment: Secondary School," The Mathematics Teacher, Vol. 68, No. 6 (October 1975), pp. 453-470.

III
Number

The discussion for this section of the report on the Year 04 NAEP mathematics assessment is based on data from approximately 100 exercises from the following NAEP content areas:

A. Number and Numeration Concepts
B. Properties of Numbers and Operations
C. Arithmetic Computation

The organization for the discussion, however, is into subsections on whole numbers, fractions, decimals and percents, integers and miscellaneous topics.

The results from all four age groups are interpreted here. It must be recognized that, to some extent, the number concepts for 9-year-olds (approximately 25 percent in grade 3, 75 percent in grade 4) are still developing, whereas the number ideas of 13-year-olds, 17-year-olds, and adults tend to have reached a terminal functional level. In other words, these exercises represent indices of achievement for 9-year-olds; for the other age groups the exercises should represent retention and maintenance of number concepts and skills as well as some degree of achievement.

For the most part, results for 9-year-olds can be interpreted in terms of whether or not the respondents have encountered the number ideas in an exercise. For the other age groups it can be assumed that the number ideas in these exercises have been covered in their mathematics curriculum.

Table 3.1 provides an analysis of the number of exercises for each content area (A, B, and C refer to the NAEP content areas) and to the exercise format—either multiple choice or open-ended. These totals refer only to the group administered exercises. The small number of individually administered items are in addition to these totals. Further, some items that will be discussed as miscellaneous topics (e.g., computation in base two) are included in these totals.

About 25 percent of the exercises are in multiple choice format. Very few of these (5) are in the arithmetic computation categories. That is, of the approximately 60 exercises in arithmetic computation, only 10 percent utilized a multiple choice format.

13

Table 3.1
Number of Parts of Multiple-Choice and
Open-ended Exercises for Number Categories

| NAEP Content: | Released | | | | | | Unreleased | | | | | |
| | A | | B | | C | | A | | B | | C | |
Exercise Format:	MC	OE	MC	OE	MC	OE	MC	OE	MC	OE	MC	OE
Whole Numbers	6	3	4	3	3	17	1	5	2	3	0	16
Fractions	1	3	3	0	0	3	0	0	0	1	2	6
Decimals and Percents	0	2	2	0	0	8	0	0	0	1	0	5
Integers	0	1	0	0	1	2	0	1	0	1	0	0

Whole Numbers

Overview of Results

Much of the early school years mathematics curriculum is devoted to learning whole number concepts, developing skill in whole number computation, and learning to use whole number concepts and operations. The Year 04 NAEP mathematics assessment included approximately 60 exercises measuring whole number topics--numeration, operations, computation, or simple problem solving. In this section, these whole number results will be examined. The following, however, are some of the major results for which supporting evidence will be presented.

I. The performance on simple numeration tasks such as recognizing place value, counting by tens, or writing numerals, was quite good for 9-year-olds.

II. The 13-year-olds and 17-year-olds performed better than adults on exercises that required respondents to show knowledge of the concept of prime numbers or to show knowledge of numeration system structure.

III. Acceptable performance was found on exercises measuring knowledge of number properties and operations.

IV. The 13-year-olds, 17-year-olds, and adults do computation tasks well. When there are differences among the three groups, the 17-year-olds and adults tend to perform at about the same level with 13-year-olds slightly lower.

V. The 9-year-olds perform well on addition problems, but poorly on the other operations.

VI. All four groups demonstrated acceptable performance on simple problem-solving tasks requiring reading (or listening to) a verbal statement of a problem, selecting a whole number operation and appropriate data from the statement, and obtaining the answer by computation.

VII. The 13-year-olds did simple problem-solving tasks far better than 9-year-olds, but less well than did 17-year-olds and adults.

VIII. The 17-year-olds and adults had similar performance levels on simple problem-solving tasks with adults slightly better on most items.

IX. With exceptions on a few exercises, the performances of all age groups on the simple problem-solving tasks were at about the same level as their performance on pure computation exercises for the same operations.

Number and Numeration Concepts

Most of the exercises calling for performance on whole number numeration topics were administered only to the 9-year-old sample. These exercises presented tasks such as recognizing the digit in the tens place of a four digit number, selecting the verbal translation of a three digit number, counting by 10's, selecting a translation from words into symbols, or determining odd or even whole numbers.

Nine-year-olds performed well on these numeration items. From 74 to 94 percent of the 9-year-olds correctly answered each exercise. Therefore, the data show a reasonably good performance on numeration concepts. There may be some disagreement with this conclusion, however, in that the tasks in the exercises on whole number numeration were quite simple. Some examples of the exercises will perhaps help the reader to judge whether the conclusion is warranted.

Table 3.2
Exercises RA01 and RA02 and Results

Exercise RA01

What digit is in the tens place in 4,263?

		9's
	No response	0
◯	2	6
◯	3	4
◯	4	8
⬤	6	75
◯	I Don't Know	7

Table 3.2 continued

Exercise RA02

762 = _____?

	9's
No response	1
○ 7 + 6 + 2	8
○ 7 + 60 + 200	5
● 700 + 60 + 2	74
○ 70 + 60 + 20	3
○ I Don't Know	8

On an unreleased open-ended exercise the 9-year-olds were asked to translate a three-digit number from words to symbols. Eighty-seven percent did it correctly.

Table 3.3
Exercise RA06 and Results

Counting by 10's, what number comes next?
10, 20, 30, _____

	9's
No response	0
40	94
20	5
Other	1
I Don't Know	0

For exercise RA06, 94 percent of the 9-year-olds responded correctly. The 5 percent who responded with "20" showed they probably understood counting by 10's but ignored or were confused by the format of the question.

Exercise RA07 (Table 3.4) was administered to both 9-year-olds and 13-year-olds. The performance level for the 9-year-olds was similar to that for exercises RA01 and RA02. Thirteen-year-olds were able to perform much better, yet almost nine percent of them failed to do the exercise correctly.

Some additional numeration exercises were administered to 13-year-olds, 17-year-olds, and adults. These exercises provide some evidence interesting in its own light, but somewhat unrelated to the above.

For example, exercise RA03 was constructed to assess place value concepts through transfer to a base 9 system (Table 3.5).

Table 3.4
Exercise RA07 and Results

Which one of the following is the sum of three hundreds, eight tens, and four ones?

		9's	13's
	No response	0	0
○	15	2	1
●	384	74	91
○	300,804	18	7
○	I Don't Know	6	1

Table 3.5
Exercise RA03 and Results

In our usual numeration system, which has 10 as its base, the numeral 25 means "2 tens plus 5 ones." In a system that has 9 as its base, the numeral 25 means

		13's	17's	Adults
	No response	0	1	1
●	2 nines plus 5 ones	42	50	38
○	2 tens plus 5 ones	30	20	17
○	5 nines plus 2 ones	5	2	4
○	5 tens plus 2 ones	2	1	1
	I Don't Know	21	27	40

The large "I don't know" response would indicate that a significant portion of each sample was unfamiliar with the idea of number bases other than 10 and elected not to try to interpret the question. For most of those who responded incorrectly to the exercise, the base 9 condition was ignored and 25 was read as a base 10 numeral. That is, they demonstrated knowledge of place value, but lack of the concept of base of a numeration system. This fact and the large "I don't know" response raises a question as to whether the concept of base, other than base 10, is covered within the school experience of most students.

One unreleased exercise asked respondents to perform a base 2 addition problem and to convert a base 10 numeral to its equivalent in base 5. The exercise was administered to 13-year-olds, 17-year-olds, and adults. Only 31 to 34 percent of the adults attempted the exercise, while 68 to 73 percent of the 13-year-olds attempted it. Very few from any age group could do either part of the exercise correctly. The results are given in Table 3.6.

To the degree that this exercise samples an acceptable goal of mathematics instruction, this performance is indeed minimal or unacceptable. The exercise indicates this material is not being learned in school programs and not being developed through incidental experiences. On the other hand, this low performance on the exercise is probably a positive sign--computation in non-base ten number bases is not very important as a goal of mathematics instruction and the low performance indicates that not much effort is devoted to it in the schools. The importance of such material would be in its use to illustrate and help students develop numeration concepts. As such, the material should represent an intermediate objective in instruction and computational facility should not necessarily be expected.

Table 3.6
Results for an Exercise in Non-base 10 Computation

		13's	17's	Adults
Addition,	No response	8	8	10
Base 2	Correct	9	8	5
	Incorrect	64	54	29
	I Don't Know	19	31	57
Convert,	No response	10	11	11
Base 10 to	Correct	18	14	6
Base 5	Incorrect	50	39	25
	I Don't Know	22	36	59

All four age groups were administered exercise RA05, a question on prime numbers (Table 3.7). The results indicate that 9-year-olds and adults did not demonstrate knowledge of prime numbers, whereas a significant fraction of the 13-year-olds and 17-year-olds correctly selected 7 as the prime number from the four numbers given.

Only eight percent of the 9-year-olds correctly selected 7 as the prime number; 25 percent of them responded "I don't know." The 9-year-olds--grades 3 and 4--are not likely to have received instruction on prime numbers. In fact, many of them will still be developing knowledge of multiplication facts at these grade levels. Adults, on the other hand, would either not have encountered prime number topics during their studies, or if they had, they would have had little opportunity to use and retain the concept after formal schooling.

Table 3.7
Exercise RA05 and Results

Which one of the following numbers is a PRIME number?

	9's	13's	17's	Adults
No response	1	1	1	0
○ 6	20	17	11	23
● 7	8	58	69	20
○ 9	9	8	9	9
○ 15	37	10	5	10
○ I Don't Know	25	6	5	38

Thirteen-year-olds and 17-year-olds, on the other hand, seem to have encountered the topic in their studies and a fair number in each age group responded to the exercise satisfactorily. The mathematics curriculum of the 1960's and 1970's would more likely have included prime number concepts than would the earlier mathematics curriculum encountered by most adults.

About the same percentage of 13-year-olds and 17-year-olds responded correctly to the prime number item (RA05) as they did to questions pertaining to even and odd numbers using symbolic representation. For example, exercise RA08 (Table 3.8) is one of the latter types of questions. Similar results (59 and 72 percent correct) were found on an unreleased exercise requiring symbolic representation of even and odd numbers.

Table 3.8
Exercise RA08 and Results

If n is an odd number, what can you say about n + 1?

	13's	17's
No response	1	1
○ It is always odd	7	5
● It is always even	51	72
○ It is even or odd depending upon what n is	39	21
○ I Don't Know	2	1

Properties of Numbers and Operations

The whole number exercises in this area, with one exception, were not administered to the adult population. That one exercise was RB06, administered to all four age groups (Table 3.9). Only a small fraction of the three oldest age groups failed to respond correctly to this exercise, indicating an operational mastery of this idea by these age groups.

Two unreleased exercises asked 9-year-olds and 13-year-olds to arrange a set of whole numbers in order from the smallest to the largest. One of the items was individually administered with each of the six numbers listed on a card. About 85 percent of the 9-year-olds and 95 percent of the 13-year-olds performed the task correctly. About 4 percent of the 9-year-olds and 3 percent of the 13-year-olds reversed the order. On the group administered exercise with four whole numbers, 86 percent of the 9-year-olds and 98 percent of the 13-year-olds did it correctly. Less than 1 percent of each age group reversed the order. As will be discussed later, both the 9-year-olds and the 13-year-olds were generally unable to perform a similar task with fractions.

Operations involving zero are felt by many mathematics educators to present special learning problems for students and to be a continual source of error in their computations. Exercise RB03 was designed to assess simple number facts involving zero. Nine-year-olds and 13-year-olds were told to do each of the problems: $3 + 0 =$ ____, $3 \times 0 =$ ____, and $3 - 0 =$ ____ (Table 3.10).

The most expected errors were 0 for the addition, 3 for the multiplication, and 0 for the subtraction. The most difficulty was found on the multiplication where 16 percent of the 9-year-olds and 5 percent of the 13-year-olds made the error of 3×0 is 3. While the percentages correct are very high, 9-year-olds should have knowledge of the generalization that any whole number multiplied by zero gives zero as the product.

Table 3.9
Exercise RB06 and Results

315 −179 136	What two numbers could you add to check this subtraction? Answer: _____ and _____			
		9's	13's	17's	Adults
179 and 136		43	83	89	86
315 and 136, or 315 and 179		9	6	3	2
Other Unacceptable		21	4	5	6
I Don't Know		24	4	3	6
No response		4	3	1	0

Table 3.10
Exercise RB03 and Results

	3 + 0		3 x 0		3 − 0	
	9's	13's	9's	13's	9's	13's
Correct	94	98	82	95	88	94
Most Expected Error	5	2	16	5	11	6
Other Error	0	0	2	0	1	0
I Don't Know	0	0	0	0	0	0
No response	0	0	0	0	0	0

Nine-year-olds were asked in exercise RB04 to select an expression equivalent to 3 x 5 (see Table 3.11). Seventy-three percent responded correctly. Since 25 percent of the 9-year-olds are in third grade, it is possible they would have minimal exposure to multiplication facts as repeated addition, although only 6 percent responded "I don't know" to the exercises. The multiple choice format would have facilitated guessing.

Comparison of two numbers, to select the one that is greater, was assessed in two parts by exercise RB02 (see Table 3.12). The first part presented two large numbers (3,000,000 and 800,000) so that comparison would have to rely on knowledge of place value. Those respondents without an understanding of place value would probably miss the item by comparing the "8" and the "3" (first digits). On the other hand, such students would be able to respond correctly to the second part of the exercise (where 3,000 and 3,200 are compared). The results indicate that only a small

Table 3.11
Exercise RB04 and Results

Which one of the following is equal to 3 x 5?

		9's
	No response	0
○	3 + 3 + 3	6
○	5 + 5 + 5 + 5 + 5	4
●	3 + 3 + 3 + 3 + 3	73
○	3 + 5 + 3 + 5 + 3 + 5	10
○	I Don't Know	6

Table 3.12
Exercise RB02 and Results

A.	Which number is GREATER?	9's	13's
	No response	0	1
●	3,000,000	82	96
○	800,000	17	3
○	I Don't Know	2	0
B.	Which number is GREATER?		
	No response	2	1
○	3,000	11	2
●	3,200	86	97
○	I Don't Know	2	1

fraction of 9-year-olds have such a limited understanding of place value in large numbers. About 4 percent of the 9-year-olds and 1 percent of the 13-year-olds made the sort of error described above while others, perhaps 11 percent of the 9-year-olds and 2 percent of the 13-year-olds were unable to correctly compare the size of large numbers. On the other hand, even allowing for guessing, the majority of the 9-year-olds could perform such comparisons quite well; 13-year-olds have the task mastered.

Only 74 percent of the 9-year-olds were able to select a true sentence connecting two unequal numerical values with the greater than (>) sign, whereas 68 percent could interpret a symbolic application of the commutative property of multiplication.

Finally, 13-year-olds and 17-year-olds were administered exercises requiring reasoning from open sentences. One such exercise is RB01 (Table 3.13). The format of the question may have influenced some of the respondents to take the "not enough information" choice. If the item had been worded "If a + 3 = b and b = 3 + c," then perhaps more students would have seen the transitivity needed to draw a conclusion. An unreleased exercise using inequalities in the item and calling for a conclusion produced slightly higher performance, 67 percent for 13-year-olds and 81 percent for 17-year-olds.

Table 3.13
Exercise RB01 and Results

If $a + 3 = b$ and $3 + c = b$, then

	13's	17's
No Response	2	0
● a equals c	57	79
○ a is less than c	5	1
○ a is greater than c	4	2
○ there is not enough information to determine the relation between a and b	27	15
○ I Don't Know	5	3

Computation with Whole Numbers

There has been considerable debate in recent years over the question of whether competence in computational skills has been declining. These national assessment data will provide no definite answer to the question. They will, however, show some examples of types of computation each age group can or cannot do well.

The following are some of the whole number computation exercises administered to all four age groups.

Exercise RC02

A. Add: B. Subtract: C. Multiply: D. Divide:

```
   38           36              38          5)125
 + 19         - 19           x  9
```

Exercise RC04

Do the following subtraction: 1,054
                                 - 865

Exercise C10076 (Individually administered)

A. Perform the addition problem...

    3 + 4 + 16 + 7 =

B. Perform the subtraction problem...

    3000 - 369 =

C. Perform the division problem...

    37604 ÷ 7 =

In addition, seven unreleased exercises were administered.
These exercises were:

1. Addition problem, two addends, no regrouping required,
    but given in verbal format: "Find the sum of a + b."

2. Addition problem, four addends.

3. Subtraction problem, in verbal format: "Subtract b from
    a."

4. Multiplication problem, two-digit by one-digit.

5. Multiplication problem, two-digit by one-digit, in
    verbal format: "Multiply a by b."

6. Multiplication problem, three-digit by three digit, with
    zero tens digit in the multiplier.

7. Division problem, three-digit by two digit.

These unreleased items were not all administered to all four
age groups. Table 3.14 presents the percent of each age group
correctly answering each whole number computation exercise.

Table 3.14
Percent Correct on Each Computation Exercise

	9's	13's	17's	Adults
RC02A	79	94	97	97
RC02B	55	89	92	92
RC02C	25	83	88	81
RC02D	15	89	93	93
RC04	27	80	89	90
C10076A	72	89	94	93
C10076B	28	71	80	87
C10076C	5	67	78	77
Unreleased Addition (1)	69	91	--	--
Unreleased Addition (2)	53	83	93	--
Unreleased Subtraction (3)	31	75	87	--
Unreleased Multiplication (4)	35	--	--	--
Unreleased Multiplication (5)	29	84	90	88
Unreleased Multiplication (6)	4	69	81	75
Unreleased Division (7)	--	66	85	--

The data for 13-year-olds, 17-year-olds, and adults do not support the view of a disaster in the area of computation skills. These data, in fact, indicate a very adequate level of performance. There is the argument of course that important aspects of computational skill (e.g., speed, mental arithmetic) were not assessed. However, the set of exercises does involve some difficult computations as well as easy ones, and the percentages of correct responses for these older age groups are good.

The older three age groups tended to be fairly close together in percent correct with the 13-year-olds slightly below the other two groups. The adult and 17-year-old performances were almost identical. On only one exercise, a subtraction problem, was the adult performance more than one percentage point above that of the 17-year-olds. Likewise, on three multiplication problems, the performance of the 17-year-olds was two to seven percentage points above the performance of the adults.

The data for the 9-year-olds indicate that computational skills have not been mastered by this age group. It is clear that dramatic increases in whole number computation skills, for all four operations, come between age 9 and age 13. The computational algorithms used in these exercises are still being learned by 9-year-olds.

Nine-year-olds could not perform the two division problems included in the assessment. That is, however, not unexpected since by grade 3 or grade 4 students will have received almost no instruction in division algorithms and will be in the process of learning multiplication facts.

Many 9-year-olds either failed to respond to multiplication or division exercises, or responded "I don't know." This further underscores that many of these respondents had not yet received instruction in the multiplication and division algorithms. Table 3.15 indicates the percent of "no response" and "I don't know" categories for the multiplication and division exercises.

The 9-year-olds' performance on the multiplication exercises of a two-digit number by a one-digit number ranges from 25 percent to 35 percent, indicating a general lack of knowledge or

Table 3.15
Percent NR or IDK Responses on Multiplication and
Division Problems for Nine-Year-Olds

	NR	IDK
RCO2C Multiplication	8	28
Unreleased Multiplication (4)	8	22
Unreleased Multiplication (5)	3	24
Unreleased Multiplication (6)	2	14
RCO2D Division	11	38
C10076C Division	14	35

skill with the multiplication algorithm. They could not perform the multiplication of a three-digit number by a three-digit number.

Each of the open-ended exercises was scored so as to detect error patterns. For example, on the subtraction problem, 36 - 19, in exercise RCO2B, 18 percent of the 9-year-olds gave 23 as the answer. They were subtracting the smaller number from the larger in each column. Twelve percent of the 9-year-olds made the same type of error on exercise RC04, 1,054 - 865. This type of error was not found in more than one percent of the other three age groups. The error indicates a lack of understanding of the nature of subtraction.

A second error pattern suggested in subtraction problems was in regrouping (or borrowing). Surprisingly, few regrouping errors were found, however, for any age group. The results for exercise RC04 illustrate this in Table 3.16. Regrouping errors were found on 10 percent of the 9-year-olds, 7 percent of the 13-year-olds, and 4 percent of the 17-year-olds and adults. Surprisingly, 14 percent of the 9-year-olds failed to respond or responded "I don't know." Thirty-six percent of the 9-year-olds' responses were scored "Other unacceptable" which may mean a larger number of errors in number combinations or it may mean some error patterns were undetected.

Table 3.16
Exercise RC04 and Results

Do the following subtraction: 1,054
− 865

	9's	13's	17's	Adults
No response	5	1	0	0
Correct	27	80	89	90
*Regrouping error at each step	1	1	0	0
*Regrouping error on one step	6	4	3	3
Added	2	0	0	0
Subtracted smaller from larger in each column	12	1	0	0
*Error with fourth digit in regrouping	3	2	1	0
Other unacceptable	36	10	6	5
I Don't Know	9	1	1	1

*Regrouping or borrowing errors.

Whole Number Word Problems

Problem solving is an integral part of mathematics learning, and the exercises presented in this section assess very elementary problem-solving skills. These exercises are all stated in some verbal context, but the background mathematical knowledge required for some of them is limited to applications of whole number computation. These exercises require primarily the translation of a verbal statement to a numerical sentence, with the corresponding choice of appropriate operations, and then completion of some whole number computation.

Certainly, other problem-solving exercises were included in the assessment. These will be discussed in other sections where the mathematical concepts required in the exercises (e.g., geometry, measurement, rational numbers) are discussed.

The following exercises are released NAEP mathematics assessment word problems requiring whole number computations for solution.

Exercise RC03

A rocket was directed at a target 525 miles south of its launching point. It landed 624 miles south of the launching point. By how many miles did it miss its target?

Answer_____

Exercise RC05

Dorothy washes windows at the rate of five minutes per window. To figure out how many minutes it will take her to wash ten windows, she could

- ○ add 5 and 10
- ○ divide 10 by 5
- ● multiply 5 by 10
- ○ subtract 5 from 10
- ○ I Don't Know

Exercise RC06

An astronaut is to orbit the earth in a space capsule for seven days. If he drinks three pints of water each day, how many pints of drinking water will he need for the trip?

Answer_____

Exercise RC07

Betty's dog eats two biscuits every day. How many days will it take the dog to finish a package of 24 biscuits?

Answer_____

Exercise RC10

Weathermen estimate that the amount of water in nine inches of snow is the same as one inch of rainfall. A certain Arctic island has an annual snowfall of 1,602 inches.

Its annual snowfall is the same as an annual rainfall
of how many inches?

Answer_____

Exercise RC11

Marie took four spelling tests. Each test had 30
words. On the four tests she spelled correctly the
following numbers of words:
25, 23, 27, and 24.
Altogether, how many words did she MISS on all
four tests?

Answer_____

Exercise RC12

A sports car owner says that he gets 22 miles per
gallon of gasoline. How many miles could he go on
seven gallons of gasoline?

Answer_____

Exercise RC13

If John drives at an average speed of 50 miles an
hour, how many hours will it take him to drive 275 miles?

Answer_____

Exercise RC19

If there are 300 calories in nine ounces of a
certain food, how many calories are there in a three-
ounce portion of that food?

Answer_____

Exercise RC21

John has 382 stamps in his stamp collection, Greg
has 224, Pete has 310 and Bob has 175. The number of
stamps the boys have altogether is CLOSEST to which one
of the following numbers?

- ⬭ 900
- ⬭ 1000
- ⬮ 1100
- ⬭ 1200
- ⬭ I Don't Know

Exercise RC23

In one year, a government department spent the following sums on four projects:
    Project A: $11,954,164
    Project B: $ 1,126,055
    Project C: $ 4,170,522
    Project D: $ 750,572
Approximately how many MILLIONS of dollars were spent on these four projects? Give your answer to the nearest MILLION dollars.

Answer_____

Table 3.17 shows the percent with correct solutions on each of these released exercises and on six unreleased exercises for each age group to which the exercise was administered. The table also indicates the whole number operation most likely utilized in the solution for each exercise.

Table 3.17
Whole Number Computation Word Problems--Percent Correct

Exercise	Operations	9's	13's	17's	Adults
RC03	Subtraction	22	72	--	--
RC05	*	50	--	--	--
RC06	Multiplication	46	--	--	--
RC07	Division	37	--	--	--
RC10	Division	--	45	72	73
RC11	Subtraction & Addition	20	66	--	--
RC12	Multiplication	--	--	89	90
RC13	Division	7	59	81	78
RC19	Division	--	--	73	81
RC21	Addition (est)	31	--	--	--
RC23	Addition (est)	--	--	54	64
Unreleased	Subtraction	76	--	--	--
Unreleased	Subtraction	41	86	--	--
Unreleased	Multiplication	69	--	--	--
Unreleased	Division	58	--	--	--
Unreleased	Division	--	39	64	--
Unreleased	Division	--	--	--	75

For exercise RC03, 22 percent of the 9-year-olds and 72 percent of the 13-year-olds solved the problem. Another 10 percent from each age group was able to write 624-525 as the appropriate operation, but gave no answer or a wrong answer. Eight percent of the 9-year-olds and 4 percent of the 13-year-olds chose to add 624 and 525.

Another characteristic error was in evidence for the 9-year-olds. Over 14 percent of them gave the result 101, which would be obtainable from 624-525 if the reversal strategy of always subtracting the smaller digit from the larger in each column was followed. Recall that 18 percent of the 9-year-olds made this reversal error on the subtraction problem in exercise RC02.

Exercise RC05 merely asked the 9-year-old respondents to select an operation, to which 50 percent responded correctly. The most illogical operation would be addition, but 22 percent of the respondents selected that distractor. This may be evidence of a predisposition to use addition in a word problem--possibly because of limited exposure to other operations in problem contexts and possibly because of greater confidence with addition facts and algorithms.

On exercise RC06, however, the most usual number sentence was $3 \times 7 = 21$. Approximately 2 percent chose to use the number sentence $7 + 3 = 10$ and 2 percent chose $7 - 3 = 4$, or attempts at either number sentence. Forty-six percent of the 9-year-olds solved the problem correctly. Compared to other performance on multiplication exercises this was very good, but the number combination 3×7 is relatively simple.

Thirty-seven percent of the 9-year-olds solved problem RC07 correctly. The most common error, other than non-pattern numerical errors, was to multiply $2 \times 24 = 48$ rather than divide. Eight percent made this error while 4 percent chose to add $24 + 2$ and 4 percent chose to subtract $24 - 2$. Eighteen percent responded "I don't know."

Exercise RC10 presented a problem requiring the division problem $1602 \div 9$ or the proportion $\frac{x}{1602} = \frac{1}{9}$ for its solution. While respective percentages correct were 31, 53, and 58, an additional 14, 19, and 15 percent of the respective age groups properly indicated the division, $1602 \div 9$. About 5 percent of each age group attempted to multiply 1602×9 to find the answer. Nine-year-olds were not administered this exercise, and respective "I don't know" percentages for the older respondents were 23, 11 and 15.

The problem presented in exercise RC11 required more than one operation, although alternative strategies were likely. For example, either subtract each of 25, 23, 27 and 24 from 30 and add the remainders, or add 25, 23, 27 and 24 in order to subtract the sum from 120. Sixty percent of the 13-year-olds solved the problem, but only 16 percent of the 9-year-olds could do it. However, another 4 percent of the 9-year-olds and another 6 percent of the 13-year-olds determined correct processes but failed to complete the solution. Nine percent and 10 percent, respectively, gave 99, the sum of 25, 23, 27 and 24, as the solution. About 20 percent and 6 percent, respectively, either did not respond or responded, "I don't know." The problem was rather difficult for 9-year-olds and provided them considerable opportunity for numerical as well as logical errors.

The sports car problem in exercise RC12 was very easy for 17-year-olds and adults. There were not sufficient errors made to determine any patterns. Likewise, RC19 was very easy for these age groups.

Exercise RC13 was administered to all four age groups. The results are summarized in Table 3.18. The solution required division of a three-digit number by a two-digit number, a process to which most 9-year-olds would not have been exposed. It is not surprising that only 7 percent were even on the way to a solution. Their general lack of background is further underscored by the fact that 18 percent of them attempted to add or subtract the two numbers provided in the problem.

Table 3.18
Results for Exercise RC13

	9's	13's	17's	Adults
No response	2	4	1	0
Correct solution	6	44	68	69
Correct process, solution not completed	1	15	13	8
Attempt to add or subtract	18	6	1	0
Attempt to multiply	2	7	2	1
Other unacceptable	42	17	11	19
I Don't Know	28	7	4	3

Exercise RC21 was structured to encourage 9-year-olds to estimate the number of stamps the four boys had altogether. It is most likely, however, that most students either guessed or carried out the actual addition. For 382 stamps, 224 stamps, 310 stamps, and 175 stamps, the total was judged closest to 900 by 18 percent, 1000 by 24 percent, 1100 by 31 percent and 1200 by 22 percent. While three column addition of four numbers would not be unfamiliar to 9-year-olds, it would be difficult. Additional information, gathered from observations of students solving this problem, would be needed to determine if 1) 9-year-olds estimate poorly, or 2) 9-year-olds tend to guess on the exercise, or 3) 9-year-olds add and make errors in doing the problem.

Exercises RC23 asked 17-year-olds and adults to estimate the number of millions of dollars, to the nearest million, from the figures given for four projects. The results are summarized in Table 3.19.

Table 3.19
Results for Exercise RC23

	17's	Adults
No response	2	0
Correct	54	64
Exact total, no estimate nor rounding to nearest million	13	11
Incorrect estimates	12	10
Other unacceptable	15	12
I Don't Know	4	4

Summary

Results indicate that 9-year-olds reached acceptable levels of performance (from 74 to 94 percent correct) on simple numeration tasks such as recognizing place value, counting by tens, and writing numerals. Older respondents did not perform as well on items designed to assess place value concepts through transfer to bases other than base 10, although these results may not be indicative of lack of knowledge of numeration concepts, but rather may indicate lack of emphasis on computational facility in non-decimal bases. A majority of 13-and 17-year-old respondents appeared to be familiar with the concept of prime number, and were also able to correctly represent even and odd numbers symbolically.

Levels of performance were high on exercises designed to assess knowledge of properties of numbers and operations on tasks such as ordering and comparing numbers and utilizing certain operational properties of zero.

The older three age groups performed well on computation tasks involving whole numbers, with acceptable responses ranging from 66 percent to 97 percent. Nine-year-olds did well on addition problems, but not as well on the other operations. These results lend credence to the argument that the current mathematics curricula have not, in fact, destroyed the development of computational skills. With few exceptions, the respondents performed as well on simple problem-solving exercises as they did on pure computational exercises.

Fractions

Overview of Results

The introduction of fraction concepts occurs early in modern school mathematics programs, but intentionally develops very gradually. By the third and fourth grades, most children should have been exposed to some informal fraction concepts, probably based on parts of a whole. To be sure, many children do have some intuitive ideas of common fractions such as 1/2, but it is not too likely that they will generally think of fractions as quantities.

In most elementary school mathematics programs today, thorough study of fraction concepts, introduction to operations with fractions, and emphasis on fractions as quantities occurs in the last half of the fourth grade, in the fifth grade, and in the sixth grade. Consequently, in terms of extant school mathematics programs, 9-year-olds (grades 3 and 4) assessed in January and February would not be expected to demonstrate much formal knowledge of fraction concepts and algorithms, although 13-year-olds should be thoroughly operational with fractions.

The following observations from the data on fractions are made. In the following pages, the evidence supporting the observations will be summarized.

I.   Only 20 to 37 percent of the 9-year-olds show knowledge of elementary fraction concepts. There appears to be an absence of fraction concepts rather than any erroneous knowledge for the majority of this age group.

II. Thirteen-year-olds and adults performed at about the same level on fraction concepts tasks, with 17-year-olds slightly better. However, the overall results were low, with at most approximately two-thirds of any age group responding correctly to an exercise that dealt with fraction concepts.

III. None of the age groups appear to possess adequate knowledge of properties of fractions and operations, especially with respect to ordering and comparing fractions.

IV. A majority of 13-and 17-year-olds could successfully multiply simple fractions (62 and 74 percent respectively) although fewer (42 and 66 percent respectively) were able to correctly add simple fractions.

Concepts of Fractions

Exercise RA10 presented figures partially shaded and asked the respondents to name the "fractional part of the figure" that was shaded. The four figures and the results for 9-year-olds are given in Table 3.20.

Table 3.20
Results for Exercise RA10

What fractional part of the figure below is shaded?

No response	2	3	3	3
Correct	31	31	31	37
Attempt to name shaded part	46	27	30	31
Ratio of shaded to unshaded	3	4	5	4
Other unacceptable	10	22	21	15
I Don't Know	7	13	10	11

From 9 to 16 percent of the 9-year-olds were unfamiliar with the material to the extent of not responding or answering "I don't know." The rather large percentage that responded to the exercise by attempting to name the shaded part (e.g., the second part of the rectangle, the left side of the circle, etc.) indicates either unfamiliarity with the content--which included the fraction concept as part of a whole and the convention of presenting fractions by diagrams such as those in the exercise--or unfamiliarity with the language (fractional part).

While diagrams similar to these are found in elementary mathematics textbooks and known to most teachers, it is unlikely that students in fourth grade would have become operational with these models of fraction concepts by mid-year. The language (fractional part), while rather self-explanatory and neutral, is not common to most elementary mathematics textbooks.

The 9-year-olds found exercise RA12, which asks for a simple application of the fraction concept, even more unfamiliar (Table 3.21). There were 54 percent of the 9-year-olds who did not respond or who responded "I don't know." Another 6 percent attempted to name the parts (e.g., "end, middle, end;" "2 sides and a middle"). This percentage is considerably lower than the similar error with the diagrams in exercise RA10. The results on exercise RA12 would argue, however, that at least 60 percent of the respondents were unfamiliar with fraction concepts in this context.

Exercise RC14 (Table 3.22) required an application of fraction concepts and was administered only to 9-year-olds. In contrast to the candy bar problem (exercise RA12) where 54 percent of the 9-year-olds were either not responding or responding "I don't know," exercise RC14 had nearly zero not responding and only 3 percent responding "I don't know." Apparently almost half of the 9-year-olds read and heard "one-fourth" to be interpreted as "one set of four," subtracted four from eight, and obtained four as an answer. This underscores a lack of understanding of fraction concepts by this age group.

Table 3.21
Exercise RA12 and Results

A candy bar is broken into 3 pieces of the same size. Each piece is what part of the candy bar?

	9's
No response	14
1/3; .33; 33%; or one-third	20
3 or 3 pieces	4
Attempt to name the parts	6
Other unacceptable	17
I Don't Know	40

Table 3.22
Exercise RC14 and Results

If one-fourth of the dots in the above figure
are removed, how many dots will be left?

	9's
*6; 6 dots; or 6 circles	21
3/4; or 3/4 of the dots	1
2; 2 dots; or 2 circles	3
4; 4 dots; or 4 circles	48
1/4; 1/4 of the dots	3
Other unacceptable	20
I Don't Know	3

One unreleased exercise also treated a fraction concept. A
small set of blue marbles and white marbles was pictured and the
respondent was asked what fractional part of the marbles was blue.
There is a fundamental difference, conceptually, between this
exercise, using discrete elements and exercises RA10 and RA12
using equal parts of a whole. The results for this unreleased
exercise are given in Table 3.23. Only 9 percent of the 9-year-
olds answered the question correctly; 36 percent gave the number
of blue marbles and 12 percent attempted to name the part of the
set that was blue (the "left side").

Table 3.23
Results on Unreleased Exercise with
Fractional Part of a Set of Discrete Elements

	9's	13's
No response	5	2
Correct	9	65
Fraction for White Marbles	–	1
Number of Blue Marbles	36	5
Ratio	3	12
Attempt to Name the Blue Part	12	—
Other unacceptable	19	13
I Don't Know	16	2

Approximately 21 percent of the 9-year-olds failed to respond or responded "I don't know." The 13-year-olds, by contrast, performed quite well on this task: 65 percent performed correctly, 12 percent named the ratio of blue marbles to white marbles, and none of them attempted to name the blue part. In short, their performance showed knowledge of fraction concepts.

Exercise RA13 was a word problem utilizing the application of fraction concepts. It was administered to 13-year-olds, 17-year-olds, and adults, and the results are given in Table 3.24.

Table 3.24
Exercise RA13 and Results

There are 13 boys and 15 girls in a group. What fractional part of the group is boys?

	13's	17's	Adults
No response	5	2	1
Correct (13/28)	20	36	25
Correct, expressed in decimal fraction or percent	--	--	1
Ratio (13/15 or .86)	17	17	15
Incorrect: Girl's fractional part	2	2	3
Other unacceptable	44	29	35
I Don't Know	12	13	21

Thirteen-year-olds and adults performed at about the same level with 17-year-olds slightly better. The results were, however, very disappointing. Approximately 15 to 17 percent of each age group made the somewhat understandable error of answering the question with the ratio of the boys group to the girls group (13/15). But 15 to 22 percent did not know how to answer the question and 31 to 46 percent made an unacceptable response other than the ratio response.

The cause of the errors cannot be determined from the data. Perhaps there is a problem with the language "fractional part" that would contribute to the "I don't know" responses. But the committed errors must be due to a lack of mastery of fraction concepts and their application to problem contexts.

Properties of Fractions and Operations

Two unreleased exercises assessed respondents' ability to
handle fractions as quantities. In one multiple choice exercise,
two very common fractions less than one were provided and 13-year-
olds were asked to select another fraction between them. Fifty-
six percent of the 13-year-olds and 83 percent of the 17-year-olds
could do the exercise correctly.

All four age groups were administered an exercise in which
six very common fractions less than one were provided and the respondents were asked to write them in order from the smallest to
the largest. No age group could perform this task adequately. The
respective percentages correct were 0.2, 3.3, 14, and 20. The
respective percentages where the error made was arranging the fractions with descending denominators, regardless of the numerators,
were 5, 18, 14, and 5. The respective percentages of other errors
where more than one pair of fractions was out of order (i.e., responses with multiple errors) were 76, 46, 26, and 25, respectively.
Hence, 17-year-olds and adults performed better both in terms of
percentages correct and fewer errors on incorrect attempts.

The descending denominators response, indicating partial knowledge of the task, was more likely to be made by 13-year-olds and
17-year-olds. This probably meant that 9-year-olds had no knowledge
appropriate to the task and adults tended to rely on more algorith-
mic strategies--like considering the decimal equivalents of the
fractions. There is, however, no reliable evidence in the exercise data to indicate strategies used in the task.

The three older age groups were asked to select a number
closest to 3/16 in a multiple choice exercise, RB08. The results
in Table 3.25 show the percentages correct were 19, 39, and 36,
respectively. The correct response was 5/32, the only alternative
with a larger denominator than the 3/16 in the stem. The most
popular choice among the distractors was 5/16 for the 13-year-olds
and 1/4 for the 17-year-olds and adults. The response of 1/4 was
second best among the alternatives, and many 17-year-olds probably
saw quite readily, or had as knowledge, that 1/4 = 4/16 and 3/8 =
6/16. Hence, from the responses easy to convert mentally to 16ths,
1/4 was the best choice. The 5/32 was probably not as familiar and
was, therefore, ignored, or an error was made in judging its "close-
ness" to 3/16. The 13-year-olds, on the other hand appear to have
preferred to select 5/16 as closest to 3/16 because it had the same
denominator. Perhaps seventh and eighth graders would not have
been as likely to mentally convert the fractions in the distractors
and the stem to a common denominator.

Table 3.25
Exercise RB08 and Results

Which number is CLOSEST to 3/16?

	13's	17's	Adults
● 5/32	19	39	36
○ 1/4	25	30	32
○ 5/16	31	15	14
○ 3/8	21	12	10
○ I Don't Know	2	4	8
No response	1	0	0

Exercise RB10 asked the respondents to select the largest fraction from among a group of four common fractions less than one (see Table 3.26). The response of 5/8 was the strongest distractor for 9-year-olds. However, the 9-year-olds are generally without adequate conception of fractions as quantity and the larger digits in this distractor undoubtedly attracted them. The strongest distractor for both the 13-year-olds and the 17-year-olds was 2/3. The exercise clearly shows that 13-year-olds are not yet operational with fractions.

Table 3.26
Exercise RB10 and Results

Which fraction is the GREATEST?

	9's	13's	17's
○ 2/3	19	44	32
○ 3/4	3	9	11
● 4/5	3	26	49
○ 5/8	71	19	8
○ I Don't Know	4	1	1
No response	0	0	0

Exercise RA11 was administered to the three youngest age groups (see Table 3.27). Nine-year-olds could not perform the task, whereas 13-year-olds and 17-year-olds gave an adequate to good performance.

Table 3.27
Exercise RA11 and Results

Which one of the following equals 47/5?

		9's	13's	17's
○	4 7/5	28	7	2
●	9 2/5	7	65	81
○	47 1/5	11	8	4
○	47 ÷ 1/5	33	14	9
○	I Don't Know	21	6	3
	No response	0	1	1

The fourth distractor, $47 \div 1/5$, was the strongest for all three age groups, possibly because it was the only distractor with an explicit operation sign.

Computation with Fractions

There were very few exercises in the assessment to measure computational facility with fractions. RC15 asked respondents to do the addition: $1/2 + 1/3 =$ _____. RC16 asked them to do the multiplication: $1/2 \times 1/4 =$ _____. An unreleased exercise, C20006, asked a question in the form: "What is a/b of c?" where a/b was a fraction less than one and c was a whole number. The results for these exercises are given in Tables 3.28, 3.29, and 3.30.

Table 3.28
Exercise RC15 and Results

$1/2 + 1/3 = ?$

	13's	17's
No response	1	1
*5/6	42	66
2/5	30	16
1/6	2	1
1/5	9	6
2/6 or 1/3	5	3
Other unacceptable	10	6
I Don't Know	1	1

Table 3.29
Exercise RC16 and Results

1/2 x 1/4 = ?

	13's	17's
No response	1	1
* 1/8	62	74
2/8 or 1/4	6	5
3/4	5	5
8	1	0
2/4; 4/8; or 1/2	8	6
1/6	4	2
Other unacceptable	11	6
I Don't Know	2	2

Table 3.30
Results for Unreleased Exercise C20006

	9's	13's	17's
No response	3	2	1
Correct Answer	17	57	78
Incorrect	53	30	15
I Don't Know	27	11	6

On the addition problem 30 percent of the 13-year-olds and 16 percent of the 17-year-olds added numerators and added denominators. This indicates a lack of understanding of fractions as quantities and a severe failure to develop computational skill or even knowledge of computation algorithms. As expected, performance on the simpler operation, multiplication, was better for both age groups.

Exercise C20006 involved a minor translation to a number sequence, selection of the multiplication operation, and multiplication of a fraction by a whole number. Seventy-eight percent of the 17-year-olds could do the task, although 6 percent said "I don't know;" 57 percent of the 13-year-olds could do the task, although 11 percent said "I don't know" and an additional 30 percent made an error. Only 17 percent of the 9-year-olds could do

the task with 27 percent responding "I don't know" and 53 percent
making errors. In view of the fraction materials present in the
third and fourth grade curriculum, the 17 percent with satisfactory
performance on this computation task is rather good.

Another exercise, RB09, called for the manipulation of an
algebraic fraction (x/y). Results of this exercise are presented
in Table 3.31. While Exercise RB09 was administered to 9-year-
olds, it is difficult to see any reasonable expectation for them
to cope with the symbolism and understand the problem. Only 21
percent responded "I don't know" and one suspects guessing played
a large part in the responses. The 13-year-olds would also be
somewhat unlikely to have encountered much algebraic symbolism,
and their performance on this exercise is undoubtedly affected by
guessing. The "double x/y" distractor is extremely compelling on
this exercise and it takes a good knowledge of fraction concepts
as quantities plus some algebraic facility to overcome the tempta-
tion to select it. The 17-year-olds were able to demonstrate reasonable performance on this exercise.

Table 3.31
Exercise RB09 and Results

Suppose x/y represents a number. If the values of x and y are
each doubled, the new number is:

	9's	13's	17's
○ 1/2 as large as x/y	17	10	8
● equal to x/y	15	18	41
○ double x/y	47	65	46
○ I Don't Know	21	7	5
No response	0	0	0

Summary

Only a small percentage of 9-year-olds exhibited any knowledge
of fraction concepts, although low results probably reflect lack of
exposure to these concepts for this age group. The older age groups
performed somewhat better, and results showed that 17-year-olds and
adults consistently outscored the 13-year-olds.

None of the age groups appeared to be knowledgeable about
properties of fractions and operations. Performances on such
exercises were consistently low, particularly with respect to
ordering and comparing fractions.

A majority of 13-and 17-year-olds were able to multiply simple fractions, and to multiply a fraction by a whole number. Much smaller percentages of these age groups were able, however, to add simple fractions with unlike denominators.

Decimals and Percents

Overview of Results

Facility in working with decimals depends heavily on knowledge of numeration and the extension of whole number computation algorithms to decimal fractions. The mathematics assessment did not include many exercises on decimals and operations with decimal fractions per se. Several exercises, especially word problems discussed in the consumer mathematics section, also assessed decimal concepts. However, there were several assessment exercises that dealt with percents; most of these exercises were in the form of word problems, and are discussed in this section. Following are some observations from the exercises on decimals and percents for which supporting data will be presented.

I.   A majority of the three older age groups were able to determine the smallest and largest decimal numerals from a given list, although selecting the largest number appears to be much easier than selecting the smallest.

II.   Expressing a given number in a different form (e.g., decimal to fraction) appears to be a difficult task for the three older age groups. Percentages of correct responses ranged from 12 to a maximum of 65 on such tasks.

III. Nine-year-olds have a knowledge and ability to do simple computation with decimals, but perform below the level of 13-year-olds. Moreover, 9-year-olds have some problem with the placement of the decimal, whereas the older age groups do not.

IV.   Basic notions of percent were generally weak for all age groups with 13-year-olds performing well below 17-year-olds and adults.

V.   In problem situations involving percents, adults performed better than 17-year-olds.

Concepts and Properties of Decimals and Percents

When asked to compare a set of positive decimal numbers to find the greatest or the smallest, 17-year-olds and adults showed a good knowledge of decimal numeration (see Table 3.32).

Table 3.32
Exercise RB07 and Results

	13's	17's	Adults
A. Which number is the GREATEST?			
○ 0.5	3	3	1
● 5.0	84	93	91
○ 0.005	12	3	3
○ 0.05	0	0	1
○ I Don't Know	1	1	3
No response	0	0	0

	13's	17's	Adults
B. Which number is the SMALLEST?			
○ 2.002	7	5	4
○ 0.202	6	6	7
○ 0.22	34	12	11
● 0.022	52	75	74
○ I Don't Know	1	1	3
No response	0	0	0

Thirteen-year-olds did well on the first part of exercise RB07, but were more prone to select a strong distractor on the second part. Most likely this distractor, 0.22, was strong for all age groups because all non-zero digits were to the right of the decimal point and there were fewer digits.

An unreleased open-ended exercise provided two decimal fractions differing by .001 and asked 13-year-olds, 17-year-olds, and adults to give a number greater than _____ but less than _____. The percentages correct were 40, 65, and 55 respectively. The "I don't know" percentages were surprisingly high: 15, 12, and 24 respectively.

Two exercises (RA14 and RA15) that measured translation from one form of notation to another were included. Exercise RA14 (Write a fraction that has the same value as .3333. . . .) had a format as shown: The ellipsis (...) followed the ".3333" and there was no separation from the period for the sentence. The exercise measured 1) knowledge of the convention for representing a repeating decimal and 2) knowledge that .3333 . . . is equal to 1/3. Almost all of the unacceptable responses had some combination of 3's in the numerator and some multiple of 10 in the denominator (e.g., 3/10, 3333/10000, .333/100, 33/100, 333/1000). The results are given in Table 3.33.

Table 3.33
Exercise RA14 and Results

	13's	17's	Adults
Write a fraction that has the same value as .3333			
No response	4	4	1
Correct (1/3)	12	41	52
3/10	3	5	1
3333/10,000	12	6	3
Other unacceptable	44	32	15
I Don't Know	26	14	28

The rather large percentages of "I don't know" responses for 13-year-olds and for adults is of concern, but the pattern of errors on this exercise indicates primarily a lack of knowledge of the convention of indicating an infinite repeating decimal with the ellipsis (. . .).

Exercise RA15, which required converting from the fraction 1/5 to its equivalent percent, should have been a recall problem for many of the 17-year-olds; knowledge of how to compute the percent given the fraction is a reasonable objective for both 13-year-olds and 17-year-olds. However, the results of this exercise, as seen in Table 3.34 show this is not the case. Less than half of the 13-year-olds and only 65 percent of the 17-year-olds responded correctly. Granted, some errors might be expected, but if the percentages are indicative of the performances of these age groups, the results are disappointing. Unfortunately, there was no parallel unreleased exercise with which to check the consistency of performance on this topic.

Table 3.34
Exercise RA15 and Results

1/5 is equivalent to what percent?

	13's	17's
No response	2	1
20 percent, or 20	41	65
.20	2	1
5 percent; 5; or .05	16	11
25, 25 percent; 25; or .25 percent	3	2
Other unacceptable	25	15
I Don't Know	11	6

Computation with Decimals

Decimal computation, in the context of dollars and cents, was assessed for all four age groups on Exercise RC01. On this straight-forward computation exercise, the 9-year-olds were much more likely to make an error in locating the decimal point (or omitting it alto-gether) as well as numerical errors. Very few made errors in carrying--indicating basic understanding of addition algorithms. Performance for the 13-year-olds and adults was below that of the 17-year-olds primarily because of their greater percentage of decimal errors.

Table 3.35
Exercise RC01 and Results

Add the following numbers: $ 3.06
10.00
9.14
5.10

	9's	13's	17's	Adults
No response	3	0	0	0
Correct	40	84	92	86
Decimal error, correct digits	22	8	2	6
Error in carrying	5	2	2	2
Other unacceptable	27	6	4	6
I Don't Know	3	0	0	0

On an unreleased exercise calling for multiplication similar to .8 x 5.6, approximately 48 percent of the 13-year-olds and 73 percent of the 17-year-olds could do the problem correctly. The errors, however, showed that 26 percent and 16 percent of the two groups misplaced the decimal as the only error.

Exercise RC17 was a subtraction problem with decimals on which all three age groups did fairly well. Very few of the errors were with misplacement of the decimal or faulty use of the subtraction algorithm.

Table 3.36
Exercise RC17 and Results

	13's	17's	Adults
If 23.8 is subtracted from 62.1, the result is:			
No response	1	0	0
Correct	61	78	74
Correct subtraction, misplaced decimal	4	3	2
Added	1	0	0
Regrouping error	3	4	5
Reversal, subtracting smaller digit from larger at each step	6	2	1
Other unacceptable	20	12	12
I Don't Know	5	2	6

Word Problems with Decimals and Percents

Exercise RC18 and RC20 presented word problems requiring knowledge of percent.

Exercise RC18

In a school election with three candidates, Joe received 120 votes, Mary received 50 votes, and George received 30 votes. What percent of the total number of votes did Joe receive?

On Exercise RC18, the respondents needed to find the total number of votes, 200, and then compute what percent of 200 is 120. Therefore, it was at least a two stage problem, requiring translation, interpretation, and computation. Approximately 18 percent of the 13-year-olds solved the problem correctly, while 45 and 48 percent of the 17-year-olds and adults were successful in solving it. Approximately 7 to 10 percent of each age group responded "I don't know." The analysis of errors, however, seems to indicate no consistent pattern, but rather a variety of numerical and procedural errors.

Exercise RC20

> Candidate A received 70 percent of the votes cast in an election. If 4,200 votes are cast in the election, how many votes did he receive?

Exercise RC20 should have been an easier word problem requiring essentially one computation, finding 70 percent of 4,200. For adults, this was the case since 62 percent responded with the correct answer, another 5 percent used the correct process but failed to complete the problem, and only 13 percent responded "I don't know."

Table 3.37
Exercise RC20 and Results

	13's	17's	Adults
No response	3	2	0
Correct result	10	41	62
Correct process, incomplete	2	3	5
Attempt to add or subtract	16	2	0
Attempt to divide 4200 ÷ .70	29	17	6
Misplaced decimal	3	2	3
Other unacceptable	17	17	12
I Don't Know	21	14	13

The problem was not as easy for 13-year-olds and 17-year-olds. Only 10 percent and 41 percent respectively answered the problem correctly. Surprisingly, 13-year-olds and 17-year-olds, 29 and 17 percent respectively, attempted to <u>divide</u> 4,200 by .70. Only 6 percent of the adults made this error. Sixteen percent of the 13-year-olds attempted to add or subtract—indicating a very inadequate knowledge of percents. In addition, 24 percent of the 13-year-olds and 16 percent of the 17-year-olds either did not respond or they responded "I don't know."

The same level of performance was found on an unreleased consumer mathematics problem where the respondents were to find the amount of a discount, expressed as a percentage, on an item in a

store. The percentages correct for 13-year-olds, 17-year-olds and
adults were 11, 41, and 73, respectively.

Summary

Certain basic concepts of decimal fractions appeared to be
well understood, although basic notions of percents were generally
weak. For example less than half of the 13-year-olds and about two-
thirds of the 17-year-olds could determine the percent equivalent
to the fraction 1/5. Furthermore only about two-thirds of the 17's
and adults knew that 100 was the comparison base for percent.
Relationships between fractions and percents as well as the comparison base for percent need to be clearly established if fundamental concepts of percent are to be effectively used. This is
clearly a mathematical topic which needs attention in both elementary and secondary school.

Computational ability with decimals appears to be fairly well-
developed, especially among the three older age groups. Nine-year-
olds could perform simple computations with decimals, although they
were more prone to make errors in decimal placement than the older
respondents.

Consumer problems involving percents were very difficult for
13-year-olds (less than 20 percent correct) and generally difficult
for 17-year-olds and adults. However, the adults consistently did
better (approximately 60 percent correct) than 17-year-olds on all
types of percentage problems. Actually the low performance by the
13-year-olds is not unexpected, since their experience with percents is limited. However, 17-year-olds should have had considerable
work with concepts of percent as well as problems involving percents.

It was also found that between 10 and 25 percent of the 17-year-
olds used a wrong operation in solving a percentage problem. For example, multiplication would be done when division was appropriate and
vice versa. Although adults made similar errors, there were fewer of
them. In most cases the use of an incorrect operation yields an un-
realistic result that could be rejected by inspection. This suggests
the possibility that 17-year-olds either did not or perhaps would not
check to see if their answer was reasonable.

Integers

Overview of Results

The few exercises that assessed computational ability with
integers were administered to the three older age groups only.
This is in keeping with the fact that integer arithmetic is
typically not included in early mathematics curricula and hence
was not an appropriate assessment topic for the 9-year-olds.

Following are some observations for which supporting data will be presented.

   I.   From two-thirds to three-fourths of the 17-year-olds could successfully perform simple computations (addition and multiplication) with integers, and from 39 to 66 percent of the 13-year-olds could do likewise.

  II.   In almost every instance, a larger percentage of both the 13-and 17-year-olds could perform a calculation involving numbers than an identical exercise expressed using letters to represent the numbers.

III.   Approximately two-thirds of the 17-year-olds and adults could perform a simple problem-solving task which involved integer arithmetic for its solution, as compared to a little more than one-third of the 13-year-olds who could perform the task.

Computations with Integers

Exercises RC24 and RC25 assessed knowledge about sums and products of two negative numbers. In the former, the 13-year-olds and 17-year-olds responded to two-open-ended problems: $(-5) + (-9) =$ _____, and $(-2) \times (-3) =$ _____. In the latter, the same age groups responded to multiple choice questions: If x and y are negative numbers, then $x + y$ is negative, is positive, or may be either positive or negative depending upon what x and y are. The parallel question was asked for x times y. These exercises and their results are presented in Tables 3.38 and 3.39.

Table 3.38
Exercise RC24 and Results

	13's	17's
$(-5) + (-9) =$ ___		
No response	0	0
-14	66	78
+14	15	10
-4	4	3
+4	6	3
-45 or +45	1	1
Other unacceptable	5	3
I Don't Know	3	2

	13's	17's
(-2) x (-3) = ___		
No response	0	1
+6	39	68
-6	48	24
-5	2	1
+5	1	1
-1	1	1
+1	2	1
Other unacceptable	4	1
I Don't Know	4	2

Table 3.39
Exercise RC25 and Results

	13's	17's
If x and y are negative numbers, then x + y		
● is negative	47	64
○ is positive	19	16
○ may be either...	30	18
○ I Don't Know	3	2
No response	1	0
If x and y are both negative numbers, then x times y		
○ is negative	29	19
● is positive	39	64
○ may be either . . .	22	11
○ I Don't Know	8	6
No response	2	0

Considering that 13-year-olds are seventh and eighth graders in the first months of the school year, many of them would not have had much exposure to integer arithmetic. Seventeen-year-olds who had experienced a non-college preparatory mathematics sequence, probably about half of them, would not likely have had any thorough exposure to integer arithmetic either. The results on Exercise RC24 and RC25 are therefore better than what might be

predicted, and the errors made are those most logical. In fact,
approximately two-thirds to three-fourths of the 17-year-olds
were successful on both exercises. As would be expected, the 13-
year-olds performed at a somewhat lower level although more than
one-third correctly responded to all portions of the exercises.

An interesting comparison may be made by examining the response patterns of the 13-year-olds to the multiplication portions
of Exercises RC24 and RC25. Although the same percentage of respondents correctly answered both questions, 51 percent of the respondents thought the product of −2 and −3 could be negative, as
compared to 29 percent on exercise RC25 who thought if x and y
are both negative, then x times y is negative. Twenty-two percent thought such a product could be either positive or negative.

Results of exercises RC24 and RC25 are paralleled by those of
exercise RC08 (Table 3.40) a simple word problem that required subtraction of a negative number for its solution. About 65 to 67 percent of the 17-year-olds and adults were able to solve the problem
correctly but only 39 percent of the 13-year-olds could solve it.
The most common error was adding the negative number--in effect not
being able to handle the subtraction algorithm for a negative number.
Again, however, this is a logical error, and is not too surprising,
given the probable lack of exposure to this topic for many of the
respondents.

Table 3.40
Exercise RC08 and Results

The air temperature on the ground is 31 degrees.
On top of a nearby mountain, the temperature is −7
degrees. How many degrees difference is there between
these two temperatures?

	13's	17's	Adults
No response	7	0	0
38; 38°; −38; −38°	39	65	67
24; 24°; −24; −24°	33	21	17
Other unacceptable	17	12	12
I Don't Know	4	3	4

Summary

   A majority of 17-year-olds were able to perform simple addition and
multiplication with integers, expressed both numerically and algebraically.
A smaller portion of 13-year-olds were able to perform these tasks, although approximately two-thirds of this age group correctly added two
negative integers. Not surprisingly, respondents were more successful
in working with numerals than symbolic letters in almost every instance.
Twice as many 17-year-olds and adults were able to solve a word problem
involving integer arithmetic than 13-year-olds.

Implications for Instruction

   The data show a mixed picture of strengths and weaknesses in
mathematics programs. Students' performance was strong or at the
level of reasonable expectation in terms of the mathematics curriculum for whole number computation, knowledge of numeration concepts,
and analysis of simple (one-step) word problems. Weaknesses in the
mathematics program of the elementary school were indicated in the
areas of percent, development of fraction concepts, and more complex word problems.

   The development of computational skills has not been destroyed
by our current mathematics curricula. The performance of 13-year-
olds and 17-year-olds is generally good on whole number computation
tasks and the performance of 9-year-olds is acceptable for grades
3 and 4. Consequently, the current retrenchment of mathematics
programs into emphasis on arithmetic skills should be examined for
finding a proper balance between skill and understanding.

   Teachers should recognize that students and adults use a variety
of approaches for solving computational problems. The data suggest
that most people use an approach that works for them in a given situation, which implies that teachers should not restrict students to
a particular approach.

   The development of problem-solving skills is extremely important;
teachers should make problem solving a regular part of mathematics
instruction. Students deserve the right kind of problem-solving experience and teachers have the central responsibility in constructing
that experience. Assessment results indicate that students apparently
receive little opportunity to learn to solve and check word problems.

   Emphasis should be placed on providing a sound initial development
of fraction concepts using concrete objects. Results indicate that
many students have little computational skill with fractions and probably little conceptual understanding. An increase in the amount of
time spent on operations with fractions is not necessarily an appropriate remedy. The development of algorithms should be paced so as to
connect firmly with the main ideas in the initial development.

<u>Experience with percent problems appears to be inadequate in current mathematics curricula.</u> In terms of funtional literacy in mathematics, 17-year-olds should be able to handle percent problems. Their low performance indicates that either the concept of percent is not well understood, or this topic is not adequately covered in mathematics curricula.

IV

Algebra

Introduction and Overview of Results

The principal focus of the algebra assessment was with 17-year-olds. Therefore, the major portion of this section will concentrate on results for 17-year-olds only, and the results for other age groups will be discussed separately.

Interpretation of the results of the algebra exercises is difficult since on many exercises respondents who had had no instruction in algebra had little chance for success. Information was not obtained on whether respondents had received instruction in algebra. Therefore, it is impossible to determine whether the low level of correct responses on most algebra exercises results from a failure to learn basic concepts or simply results from lack of exposure. Similar dangers are encountered in analyzing errors.

With the preceding discussion in mind, following are the major observations that may be made.

I. Approximately half of the 17-year-olds were able to translate verbal statements into algebraic expressions.

II. A majority of 17-year-olds were able to evaluate algebraic expressions in one variable, but were unsuccessful for the most part in evaluating algebraic expressions in more than one variable.

III. The 17-year-olds were generally successful in solving simple linear equations, but only about one-third could solve equations that required several steps in their solution.

IV. Approximately one-fifth of the 17-year-olds could solve a system of linear equations, factor quadratic expressions, solve simple quadratic equations, and graph linear equations.

V. Ninety percent of the 9-year-olds could successfully solve an addition open sentence, but only about half of this age group could successfully solve an open sentence involving subtraction.

VI. Most 13-year-olds were successful in solving an open sentence whose solution could be found by inspection, but fewer than half could solve an equation that required several steps for its solution.

57

VII. About half of the 13-year-olds were successful
      in evaluating simple algebraic expressions for specific
      values of the variable.

VIII. Adults scored consistently lower than 17-year-olds on
      algebra exercises; scores of the adults were generally
      close to those of the 13-year-olds.

Selected Topics of Algebra

Constructing Algebraic Expressions

Applying algebra to problem situations is one of the fundamental objectives of the algebra curriculum. A basic step in such
application is translating verbal problems to algebraic expressions.
It has long been the contention of many mathematics teachers that
solving verbal problems that involve such translations is one of
the most difficult algebra topics and is the topic with which pupils
are least successful. This contention is not supported by the
National Assessment data. The results of three exercises summarized
in Table 4.1 indicate that respondents were more successful translating verbal problems to algebraic expressions than they were with
most other algebra topics. This conclusion must be tempered by
several factors. First, the problems were not difficult, the variables for the algebraic expressions were actually contained in the
verbal statements, and they did not involve equations. Second, unlike most other algebra topics, the exercises did not necessarily
require any specific instruction in algebra and may have been solved
correctly by respondents who had not taken an algebra course. In
any case, the results indicate that pupils may have greater success
translating verbal statements to algebraic expressions than most
teachers suspect.

Table 4.1
Exercises RG01, RG02, G20002 and Results

RG01 If y dollars are shared equally among four boys, how many
      dollars does each boy receive?

RG02 Apples cost a cents per pound and pears cost p cents per
      pound. What is the total cost of two pounds of apples and one
      pound of pears?

G20002 (Unreleased) Addition and subtraction

	13's	17's	Adults
RG01	12	43	33
RG02	21	51	--
G20002	20	44	--

Evaluating Algebraic Expressions

Evaluating algebraic expressions for specific values of the variables is a basic process that is used in applying formulas, solving systems of equations, graphing equations, and checking solutions to equations. The results summarized in Table 4.2 indicate that pupils generally understand the concept of substituting specific values for variables and can evaluate simple expressions involving either equalities or inequalities. However, respondents had significantly greater difficulty with equations in

Table 4.2
Exercises RG04, RH05, RH04, RH07, RI01 and Results

RG04   If x = 3, what is the value of $x^2 - 1$?

RH05   If x is less than 4, then x + 7 MUST be
○ less than 7
● less than 11
○ greater than 7
○ greater than 11
○ I Don't Know

RH04   Given the equation, 2x + 3y = 12, complete the table below.

x	-3		0
y		4	

RH07   The formula for the relationship between Fahrenheit and Centigrade temperatures is $F = \frac{9}{5} C + 32^o$. What is C when F is 77° ?

RI01   If f(x) = x + 1, what is the value of f(2)?

	13's	17's	Adults
RG04	45	75	48
RH05	50	70	56
RH04	1	24	--
RH07	2	24	--
RI01	--	40	--

---

several variables, even with relatively simple equations. In RH04,
less than 40 percent of the 17-year-olds were able to find any of
the corresponding values of x and y and only 24 percent found all
three. In addition to substituting values in an expression, this
exercise required solving an equation. However, the results for
RH02 (see Table 4.3) indicate that 75 percent of the 17-year-olds
can solve similar equations. Thus, over 75 percent of the 17-year-
olds have mastered the prerequisites for this type of problem but
over half are unable to integrate them.

Table 4.3
Exercises RH02, RH03, H11007, H11015 and Results

RH02   What is the value of x that satisfies the equation
3x - 3 = 12?

RH03   What value of x makes the following TRUE? x - 3 = 7

H11007 (unreleased)

Equation of the form $\frac{a}{b} = \frac{x}{c}$ where a, b, and c are

whole numbers and c is a multiple of b.

H11015 (unreleased)

Equation which requires combining terms. Both
numbers and variables appear on both sides of equal
sign. Whole number solution.

	9's	13's	17's	Adults
RH02	--	39	75	--
RH03	49	85	95	81
H11007	--	64	77	56
H11015	--	--	36	25

Many respondents who were able to evaluate simple expressions
appeared to be unfamiliar with function notation (see RI01).

Solving Single Value Linear Equations

Seventeen-year-olds were generally successful solving equations
as long as the solution could be found by observation (see RH02,
RH03, and H11007 in Table 4.3), but only 36 percent could solve an
equation that required several steps for solution (H11015).

Systems of Equations

Only one unreleased exercise tested the ability of 17-year-olds to solve systems of equations. Seventeen percent correctly found the solution for a pair of linear equations in two unknowns. This is a slightly smaller proportion than those who could substitute numerical values into a linear equation in several variables. (See Table 4.2, RH04.) A pupil who cannot correctly substitute numerical values into an equation of several variables is a long way from being able to apply the substitution method of solving systems of equations and would most likely have difficulty with any other method. Thus, it appears that pupils either can successfully solve systems of equations or have not mastered many of the prerequisites. There is very little middle ground.

Quadratic Equations

As with systems of linear equations, the ability to solve quadratic equations is very much an all-or-none situation. Either pupils can solve quadratic equations by factoring or they have not mastered many of the prerequisites. The results for four exercises dealing with concepts relating to quadratic equations are summarized in Table 4.4. Virtually all the respondents who could factor the expression $x^2 - bx + c$ could solve the equation $x^2 - bx + c = 0$. By the same token, finding the solution to a factored quadratic equation (RH06) was just slightly easier than solving the unfactored quadratic.

Table 4.4
Results for Exercise RH06 and Three Unreleased Exercises

RH06	What is the solution set of the equation $(x-1)(x+7) = 0$?
G43005	(unreleased)	Multiplication of binomials
G44007	(unreleased)	Factor expression of the form $x^2 - dx + e$, d, e, whole numbers
H22005	(unreleased)	Quadratic equation of the form $x^2 - dx + e = 0$ same d, e, as in G44007

	17's
RH06	26
G43005	30
G44007	20
H22005	19

Graphing Equations

Seventeen-year-olds were able to graph linear equations with about the same degree of success that they had with other non-trivial algebra exercises. However, exercises dealing with slope and intercept and equations of circles were among the most difficult exercises administered.

As with the systems of equations, about the same proportion of respondents that could substitute numerical values into a linear equation (see Table 4.2, Exercise RH04) could find the coordinates of specific points of a line and correctly graphed the line (20 percent correct). Identifying the equation of a horizontal line was of about equal difficulty (21 percent correct), but finding the slope and y-intercept of a line was somewhat more difficult (16 and 12 percent correct respectively), and finding the equation of a line given its x and y coordinates was more difficult still (5 percent correct). The most difficult graphing exercise was to find the equation of a circle with a given radius and center at the origin (3 percent correct).

Summary

It appears that the knowledge of basic algebra of 17-year-olds is very much an all-or-none situation. With the exception of simple exercises that could be solved by inspection, about the same proportion of 17-year-olds could successfully apply most basic algebra processes tested. Exercises testing graphing equations, solving systems of equations, and solving quadratic equations were all solved by about a quarter of the 17-year-olds. Furthermore, this consistency also applied to hierarchical collections of skills. Respondents who could not solve a given exercise generally were unsuccessful with exercises testing prerequisite skills. For example, the proportion of respondents who could solve an unfactored quadratic equation was about the same as could factor the quadratic expression or could solve a factored quadratic equation. Similarly, about the same proportion that could successfully substitute numerical values into a linear equation to solve for the other variable could graph linear equations and solve systems of equations.

On the other hand, a significantly greater proportion demonstrated some ability to deal with the concept of a variable and the abstract symbolism of algebra. About half of the 17-year-olds could translate verbal problems to algebraic expressions, and about three quarters could solve simple equations and evaluate expressions for specific values of the variable.

It appears as though the distinguishing factor is whether specific, well-defined procedures are required for the solution of the problems. When they are, about a quarter of the 17-year-olds can solve the problem. The rate of success is significantly greater for problems that can be solved intuitively.

Results for 9-Year-Olds

Four exercises, two of which are released, tested 9-year-olds' ability to solve simple equations or open sentences. The results indicate that 9-year-olds can successfully operate with simple open sentences, as 90 percent successfully solved the addition exercise with open sentences. The significant increase in difficulty for the subtraction exercise is probably the result of several factors. Subtraction sentences may be inherently more difficult. Only 2 percent of the 9-year-olds used the incorrect operation in the addition exercises whereas 13 percent did in the subtraction exercise. It is probably also true that pupils have considerably more experience with addition open sentences than with subtraction. Addition sentences are used to motivate subtraction, but there is no comparable use of subtraction open sentences. On the other hand, much of the difficulty may have resulted from the fact that the subtraction exercises employed the variable "x" rather than the box. Nine-year-olds are probably more familiar with the box notation, as it is commonly used in primary texts whereas the "x" is not. The fact that 20 percent more respondents failed to respond or chose the "I don't know" response when the "x" was used offers some support for this explanation.

Performance on unreleased multiplication and division sentences corresponds favorably to the results of exercises testing computation with multiplication and division.

Table 4.5
Exercises RH01 and RH03 and Results

		Percent Giving Response 9's
RH01	Replace the box with a number to make the following statement TRUE. $3 + \square = 10.$	
RH03	What value of x makes the following TRUE? $x - 3 = 7$	
RH01		
	Correct	90
	13	2
	Other Unacceptable	5
	No Response or I Don't Know	3
RH03		
	Correct	49
	4	13
	Other Unacceptable	15
	No Response or I Don't Know	23

Results for 13-Year-Olds

Few 13-year-olds have had a formal course in algebra. Therefore, it is to be expected that their experience with algebra concepts is limited. The results for RH03 (Table 4.3) indicate that most 13-year-olds can solve simple equations that can be solved by inspection. Exercise RH02 (Table 4.3) which required several steps for solution, was too difficult for the majority of 13-year-olds as 30 percent were unable to respond and another 31 percent gave incorrect responses. However, considering that few 13-year-olds have much experience with such equations, it is not surprising that the majority cannot solve them; and it is a positive sign that close to 40 percent could generate a solution.

The results in Table 4.2 indicate that about half the 13-year-olds were able to evaluate simple algebraic expressions for specific values of the variable. The results for RH05 appear to reflect difficulty in dealing with variables within inequalities. However, the results for RG04 are undoubtedly affected by 13-year-olds' unfamiliarity with exponential notation. Only 45 percent of the 13-year-olds were able to evaluate 3^2. This is reflected in the fact that close to 20 percent made errors that were clearly based on a misinterpretation of the exponential notation (ignoring the exponent, multiplying the 2 times the 3). Therefore, as far as indicating the ability of 13-year-olds to evaluate expressions for numerical values of the variable, these results are somewhat on the low side.

Most other algebra exercises were correctly solved by fewer than one-fourth of the 13-year-olds. Between 10 and 20 percent correctly translated verbal statements to algebraic expressions, only 13 percent could identify the graph of $y = x$, and only 19 percent could graph a point in the first quadrant of a coordinate system.

Results for Adults

Adults were not tested for exercises that required a sequence of specific procedures for solution, like solving systems of equations, solving quadratic equations, etc. Their performace was uniformly below that of 17-year-olds on all exercises administered to both groups with the difference generally falling between 10 and 20 percent. In fact, adults and 13-year-olds generally scored within a few percentage points of one another, except that adults were significantly more successful in translating verbal statements to algebraic expressions (RG01, Table 4.1).

Implications for Instruction

The lack of information on the number (or proportion) of respondents who had taken algebra makes interpretation of the

results difficult. Any inferences drawn from the data must be
tempered by this fact. Suffice it to say that there is plenty of
room for improvement. Furthermore, teachers should be reminded that
mathematics instruction from grades seven to nine is extremely important for developing functional algebraic skills.

V

Geometry

Introduction and Overview of Results

    There were 20 released and 23 unreleased exercises originally
classified as geometry. However, since several of these exercises
deal primarily with measurement, e.g., length, perimeter, area, and
volume, they are discussed in the chapter on measurement. The observations and interpretations given in this section are therefore based on
results from the remaining 16 released and 15 unreleased geometry
exercises. The distribution of these exercises across content topics
and age levels is given in Table 5.1. It should be noted that unreleased Exercise K30027 consists of several sections involving the
recognition of two-and three-dimensional figures. It is counted as
one exercise at each age level in Table 5.1. In other discussions
the various parts of the exercise are recognized.

Table 5.1
Distribution of Geometry Exercises

Subtopics	9	13	17	Adult	Exercises
Recognition of Geometric Shapes	11	6	4	3	13
Applying Geometric Relationships	7	10	13	7	18
Totals	18	16	17	10	31

    The exercises represent an adequate sampling of some of the key
ideas of geometry: recognition of plane and solid figures, including
diameters of circles, angles, and parallel lines; and applying certain
basic principles and concepts involving angles, circles, and squares.
Discussion of the exercises will follow the breakdown presented in
Table 5.1.

Recognition of Geometric Shapes

Overview of Results

Following are the major observations for the 13 exercises that dealt with recognition of geometric shapes.

I.   Nine-year-olds performed at a near satisfactory level at recognizing and naming certain basic plane shapes, e.g., circle, rectangle, and triangle. The range of acceptable responses for this type of exercise was from 72 percent to 96 percent.

II.   The majority of 9-year-olds are not familiar with basic relationships or the vocabulary concerning diameter, parallel lines, or right angles.

III. On those exercises administered to both 9-and 13-year-olds, the 13-year-olds uniformly performed at a substantially higher level, usually from 25 to 40 percent higher; although the performance for 13-year-olds could not in general be classified as satisfactory.

IV.   Recalling the name of a geometric solid from memory when looking at a model is more difficult than identifying the name of the shape from a list.

V.   Thirteen-year-olds are not familiar with the names of certain common solid figures--especially cylinder, cube, and sphere. It seems clear from the results that 13-year-olds have received some instruction with solid shapes but apparently the technical names for these shapes have not been emphasized.

VI.   The use of everyday vocabulary to name solid shapes persists at a fairly constant level from 9-year-olds to adults.

VII. As the level of performance for naming solid shapes increases, the frequency of using names of related plane shapes decreases and the frequency of using length, width, and height terminology in explaining the difference between plane and solid shapes increases.

VIII. Adults perform below the level of 17-year-olds on most geometry exercises. This is contrary to the general pattern shown by these two age groups on measurement, computation, and consumer mathematics exercises. One exception involves the naming of a cube on which the adults scored 11 percent higher than the 17-year-olds.

Lines and Angles

   Parallel lines. There were two released exercises dealing with parallel lines. One of these (RK02) involved the actual drawing of a line through a given point and parallel to a given line. Exercise RK01 was a multiple-choice exercise administered only to 9-year-olds in which the respondents were to select the picture showing parallel lines. The results indicate that only 48 percent of 9-year-olds were able to identify a picture showing parallel lines with 28 percent responding "I don't know." Exercises RK01 and RK02, and results, are presented in Table 5.2.

Table 5.2
Exercises RK01 and RK02 and Results

RK01: Which picture shows parallel lines?

		9's
●	↔ ↔	48
○	⤬	8
○	↑→	7
○	⤢↔	7
○	I Don't Know	28
	No response	1

RK02: Draw a line through point P that is parallel to L. Use the ruler you have been given.

L _____

.P

	9's
Acceptable Response	18
Unacceptable Response	73
I Don't Know	5
No response	4

The drawn line in Exercise RK02 had to touch or nearly touch line L to be scored as unacceptable. Fifty-six percent drew the line in this manner with an additional 10 percent drawing it almost perpendicular to line L. Skill in performing the drawing was a negligible factor. Nearly 2 percent of the responses were scored as acceptable even though the line did not pass through point P.

The term "parallel" does not appear to be in the vocabulary of 9-year-olds. Although it is more difficult to produce a drawing illustrating parallel lines than it is to identify a picture of parallel lines, the results for both of these exercises are disappointing. The concept and terminology of parallel lines is not only important to the development of subsequent mathematical ideas, it is also useful in everyday life. The low level of recognition of a picture of parallel lines is probably more indicative of a lack of instruction concerning parallel lines in the primary grades than any inherent difficulty in the concept of parallel lines.

Right angles and angle measures. There were three released exercises dealing with the recognition of a right angle and the degree as a unit of angle measure. RK03 was administered to 9-year-olds only, while RK04 was given to 13-year-olds and 17-year-olds. RE01, a measurement exercise, was administered to each age group. These exercises, and results are presented in Table 5.3, 5.4, and 5.5.

Table 5.3
Exercise RK03 and Results

	Angle A is what kind of angle?
A	
	9's
○ Acute	15
○ Obtuse	19
● Right	32
○ I Don't Know	34

Table 5.4
Exercise RK04 and Results

What is the measure in degrees of the angle formed by the hands
of the clock when the time is three o'clock?

	13's	17's
Acceptable Response	43	73
Unacceptable Response	34	20
I Don't Know	17	6
No response	6	1

Table 5.5
Exercise RE01 and Results

An angle may be measured in units called:

	9's	13's	17's	Adults
○ centimeters	18	10	5	4
● degrees	15	69	89	82
○ grams	10	3	1	1
○ inches	43	15	4	7
○ I Don't Know	14	3	1	6

It is not surprising that 9-year-olds did not correctly name the
angle in Exercise RK03 or know the term "degrees" as shown by Table 5.5.
The concept and the terminology of angles are not developed in most
schools during the primary years. It is of concern, however, that only
43 percent of the 13-year-olds responded correctly to Exercise RK04
(Table 5.4). Among several possible approaches to the solution of this
exercise, a likely one would have the respondent visualize that the
angle formed by the hands at 3:00 is a right angle and recall that the
measure of a right angle is 90 degrees (270 degrees was also acceptable).

An unreleased exercise, K11004, tested 9-and 13-year-olds' knowledge of the number of degrees in a right angle. Only 4 percent of the
9-year-olds responded correctly. Fifty-five percent of the 13-year-
olds responded correctly while 20 percent indicated that they did not
know. From this it seems apparent that the 13-year-old respondents to
Exercise RK04 had less difficulty telling how many degrees in a right
angle than they did in visualizing the right angle from the position
of the hands on the clock.

The results from Exercise RE01 support the interpretation that 9-year-olds are unfamiliar with angle measure. While the performance level for 13-year-olds was higher than for 9-year-olds, it is of concern to note that 15 percent selected "inches" as units of measure for angles. A satisfactory percentage of 17-year-olds and adults seem to know that angles may be measured in degrees.

The results on Exercise RK04 for 17-year-olds are slightly less than satisfactory. If it can be assumed that a 17-year-old should be able to visualize the hands of a clock at 3:00, then it appears that about one-fourth of the 17-year-olds either do not know that there are 90 degrees in a right angle or that there are 360 degrees in a circle. Relatively low performances appear to be the result of a lack of appropriate instruction or inadequate attention given to the initial development of the idea during the early secondary school level.

Two-dimensional Shapes

Rectangles and triangles. There were two released exercises and two unreleased exercises that involved recognition of two-dimensional geometric shapes. Each of the two released items was administered to 9-year-olds only. These exercises and the results are given in Tables 5.6 and 5.7.

Table 5.6
Exercise RK05 and Results

Which one of the following figures is a rectangle?

	9's
○ ▱	3
● ▭	74
○ ⏢	2
○ △	20
○ I Don't Know	1

Table 5.7
Exercise RK09 and Results

What is this figure called?

9's

Acceptable Response	72
Unacceptable Response	11
I Don't Know	14
No response	3

Nine-year-olds appear to be slightly confused in their recognition of a rectangle. While 74 percent correctly identified the picture of a rectangle in RK05, 20 percent selected the picture of a triangle. This confusion might result from the similarity in the ending sounds of the words rectangle and triangle; or it may be that the triangle was a popular alternative since it was the only non-quadrilateral given. Interestingly, the name rectangle was not frequenlty confused with the name triangle in exercise RK09 (Table 5.7). In fact, fewer than three percent of the respondents labeled the triangle as a rectangle.

Unreleased Exercise K20050 required the recognition of a specified regular polygon, given the name and four drawings to select from in a multiple-choice format. This exercise was administered to all but 9-year-olds. Fifty-eight percent of the 13-year-olds recognized the name of the given shape and were able to identify the drawing as compared to 84 percent of the 17-year-olds who were successful on this exercise. The results for young adults may indicate that they do not maintain the recognition of this particular regular polygon, as is evidenced by a 10 percent reduction in performance level from that of the 17-year-olds.

Exercise K30027 was an individually administered unreleased exercise the focus of which was to determine the extent to which each of the four age groups could differentiate between two- and three-dimensional figures, although a portion of the exercise was concerned with the naming of geometric figures. Eight-eight percent of the 9-year-olds were able to name a triangle when shown a concrete model in this situation. This level of performance is substantially higher than that for Exercise RK09.

This result is confounded, however, by the fact that due to the nature of the exercise, the majority of respondents had a time interval during which they were classifying and sorting the concrete models prior to naming them. This extra activity may have enhanced the recall for many of the respondents.

The exercises reported above were considerably different in
format and were not administered to the same subjects, making it impossible to make too many comparisons between them. Suffice it to say
that about one-fourth of the 9-year-olds did not recognize a picture
of a rectangle or could not name a triangle from a picture, although
a substantially greater percentage did say the name triangle when shown
a concrete model. While the latter result is somewhat more satisfactory,
the overall results seem to indicate a lack of appropriate attention
given to geometric shapes and vocabulary in the primary grades.

Circles and Diameters

There were no released exercises involving the recognition or
naming of a circle. Individually administered Exercise K30027 did contain a task (previously described in this section) in which the respondents were to name a concrete model of a circle. Ninty-six percent of
the 9-year-olds were able to do this, as were 95 percent of the 13-
year-olds, 97 percent of the 17-year-olds and 91 percent of the adults.

The two released exercises involving circles actually focused on
naming and recognizing the diameter. These exercises and the results
are given in Tables 5.8 and 5.9.

Table 5.8
Exercise RK06 and Results

Which line segment is a DIAMETER?

		9's	13's
○	EG	6	3
○	HK	14	5
●	HM	28	68
○	NP	26	17
○	I Don't Know	26	7

Table 5.9
Exercise RK10 and Results

What is line segment PT called?

		9's
⬭	centerline	45
⬭	diagonal	10
⬬	diameter	17
⬭	radius	8
⬭	I Don't Know	19

Although 9-year-olds could name a circle from a model, they were not sufficiently familiar with the parts of a circle to be able to name or recognize the diameter. This is primarily due to the fact that little or no instruction is given during the primary grades to this topic. It is interesting to note on Exercise RK10 that 45 percent selected "centerline" as the name for segment PT. This result probably indicates that many 9-year-olds know something about the center of a circle and quite naturally viewed PT as a "centerline."

The results for 13-year-olds on Exercise RK06 are disappointingly low. The typical 13-year-old has completed the elementary school mathematics program. The fact that only 68 percent could recognize the diameter of the given circle is indicative of a lack of appropriate instruction and a lack of attention given to the explicit use of the term diameter.

Three-dimensional Shapes

There was only one released exercise involving the recognition of three-dimensional figures. Three unreleased exercises, two of which were individually administered, also involved naming or recognizing three-dimensional figures. The vocabulary of common solids such as cylinder, cube, sphere, and cone received some emphasis in these exercises.

Exercise RK11 was administered to 9-year-olds only. It attempted to measure recognition of sphere as the name of the shape most like an orange. Knowledge of vocabulary was essential here since the respondent must select the word from the given list.

Table 5.10
Exercise RK11 and Results

Which one of the following has a shape <u>MOST</u> like an orange?

		9's
○	Cone	14
○	Cube	14
○	Cylinder	36
●	Sphere	24
○	I Don't Know	11

In most schools, 9-year-olds have not formally studied the sphere. One would assume, however, that given a collection of models for the solids, including a sphere, the majority of 9-year-olds would make a correct identification when asked, "Which of these (pointing to the collection of choices) has a shape most like an orange?" Difficulty with this exercise apparently resulted from unfamiliarity with the term "sphere." The results indicate that the respondents may have been guessing between sphere and cylinder, with more choosing cylinder.

In unreleased Exercise K20037 the respondents were to identify the names of three solids, each resembling everyday objects. This multiple-choice exercise, which was similar to RK11, was administered to each age group. Again, the low results for 9-year-olds shown in Table 5.11 indicate the lack of formal teaching of the vocabulary of these common solid figures in the primary grades. The increase in performance between 9-year-olds and 13-year-olds is greater than that between 13-year-olds and 17-year-olds. It is probable that most students learn the names of these solids between grades 4 and 8 although learning continues into secondary school. The higher performance for rectangular solid may be due to the fact that it was the only two-word choice given and each word is strongly suggestive of the familiar object stated in the exercise.

Table 5.11
Percent Selecting Name of Solid Representing
Shape of Familiar Object
(K20037)

Correct Response	9's	13's	17's	Adults
Rectangular Solid	49	82	92	86
Cylinder	21	66	82	81
Sphere	29	69	84	73

As in Exercise RK11, the most frequent incorrect choice for each
age group on the sphere question was the cylinder response and vice-
versa for the cylinder question. Consider, for example, results for
the 13-year-olds. Sixty-six percent correctly selected cylinder with
16 percent choosing sphere, whereas 69 percent were correct on the sphere
identification with 10 percent selecting cylinder.

A portion of Exercise K30027 (discussed in a previous section) was
concerned with the naming of certain solid figures as plastic models
were held up by the examiner. Results from this portion of the exercise
are presented in Table 5.12.

Table 5.12
Percent of Acceptable Responses for
Naming a Model of a Solid
(K30027)

Solid Figure	9's	13's	17's	Adults
Cone	28	54	74	72
Cylinder	3	24	53	56
Cube	4	26	43	54
Sphere	2	21	46	41

Results indicate that of the cube, sphere, cone and cylinder, the
cube, sphere, and cylinder were difficult for each age group, while the
cone was the most often correctly named object. This high level of
performance may be because the mathematical name for a cone is also one
of the most common everyday names for this shape.

Naming the actual model of the solid was more difficult than selecting the name from a list of names when given a written suggestion of the solid. For example, only 46 percent of the 17-year-olds named the sphere from memory in K30027; whereas 84 percent selected the word sphere, after having it read on the audio tape, as the shape suggested by a stated familiar object in K20037. It is understandable that hearing and recognizing the word is less difficult than recalling it from memory.

There appear to be two factors related to interference in learning the names of certain solid figures. One factor involves what might be called everyday vocabulary for the solids, and the other involves the names of related two-dimensional figures. Table 5.13 indicates that the percentage of each age group that selected either an everyday vocabulary term or a two-dimensional name for each solid.

Table 5.13
Percent Giving Unacceptable Responses
For Naming a Model of a Solid
(K30027)

Solid Figure		9's	13's	17's	Adults
Cone:	Everyday vocabulary	7	6	4	6
	2-Dimensional	19	5	2	1
Cylinder:	Everyday vocabulary	20	16	11	12
	2-Dimensional	28	8	4	4
Cube:	Everyday vocabulary	12	14	13	11
	2-Dimensional	81	58	41	32
Sphere:	Everyday vocabulary	37	44	33	40
	2-Dimensional	49	24	13	9

The use of everyday names persists at a fairly constant level, especially high for the sphere , in each of the age groups. The frequency of such names as ball, globe, etc., was highest for 13-year-olds. This may indicate that the word sphere is not being widely stressed when the sphere and its properties are studied during the later elementary grades.

The data on the frequency of using 2-dimensional names for each solid indicates that this usage drops off as age level increases. This may indicate that instruction successfully counts some of the initial two-dimensional perceptions of these solids, yet does not stress formal names enough to counter the influence of everyday vocabulary. Knowledge of the difference between plane and solid figures was also assessed in Exercise K30027. A verbal differentiation between plane and solid figures was a difficult task for all age levels, with levels of acceptable performance ranging from 9 percent correct to 50 percent correct for 17-year-olds. Levels of success were much higher on the task of selecting specified two-and three-dimensional models from a collection of models.

Summary

Most 9-year-olds could recognize and name basic plane shapes such as circle, rectangle, and triangle, but few respondents in this age group were familiar with the terminology for diameter of a circle, parallel lines, and right angle. Thirteen-year-olds were generally more proficient in recognizing and naming diameters, parallel lines, and right angles, although performance was far from perfect.

Nine-tenths of the 9-year-olds and almost two-thirds of the 13-year-olds could not adequately explain the difference between plane and solid figures. Although two-thirds of the 13-year-olds could recognize such common solid figures as a cylinder or a sphere, only one-fourth could recall the names of these figures. The majority of 17-year-olds and adults could distinguish between plane and solid figures and recognize the names of some of the common solid figures, but only about half of them could recall the names of the cube, cylinder, and sphere.

Applying Geometric Relationships

Certain exercises required knowledge of a basic geometric concept or the application of a geometric principle or relationship in order to solve a problem. In general, these exercises were more difficult than those involving recognition of a shape or the naming of a figure. Following are the major observations for which supporting data will be presented.

I.   Properties of a square could be used to solve problems by a majority of the 13-year-olds but by only a few 9-year-olds.

II.   Thirteen-year-olds, 17-year-olds, and adults are not adequately familiar with angle relationships other than right angles.

III.   Properties of circles, radii, diameters, and central
       angles were not known to 9-year-olds and 13-year-olds
       showed unsatisfactory performance on exercises about
       circles.

IV.   Seventeen-year-olds are more successful at solving
       certain geometric problems than 13-year-olds, but the
       majority are not successful at solving problems
       involving proportions or the Pythagorean theorem.

Plane Figures

Squares.   There were exercises, one released and two unreleased,
that required application of properties of a square for their solution.
The first of these, Exercise RK16, was administered to 9-and 13-year-
olds.

Table 5.14
Exercise RK16 and Results

Shown below are two squares.   A and B are the centers
of the squares.   What is the distance in inches from
A to B?

	9's	13's
Acceptable Response	36	60
Unacceptable Response	43	31
I Don't Know	19	6
No response	2	2

This exercise appears to be too complex for the majority of 9-year-
olds.   Other results on geometry exercises have indicates that 9-year-
olds are just mastering recognition of shapes and are not ready to deal
with problems involving properties of those shapes.

Twenty-four percent of the 13-year-olds answered 4 or 4 inches,
which was the majority of unacceptable responses given by the 13-year-
olds.   Only 17 percent of the 9-year-olds gave this response.   More of
the 9-year-olds simply indicated that they did not know.   "Four" is the
answer that a student could give in an attempt to solve the problem but
yet be unable to make the crucial translation of the given dimensions
to the horizontal segments.

In unreleased Exercise K20008, the respondents were to determine the length of one side of a square, having been given the length of one of the other sides. This open-ended exercise was administered to 9-year-olds, and 67 percent gave an acceptable response while 8 percent indicated that they did not know. A drawing of a square was used and the length of one side was clearly labeled. The results may be somewhat artificially high in the sense that a respondent would not have to know the essential property of a square in order to get a correct answer--the only number given in the problem is the correct answer.

The results from K20008 indicate that 67 percent of the 9-year-olds can translate the labeled dimension to another side of a square. A similar unreleased exercise, K10010, required that the respondents take half of the length of one side of a square. The problem was complicated somewhat by the fact that the segment whose length was given was parallel to a side and separated the square into halves. The word parallel is a proven distractor. This exercise was administered to 13-and 17-year-olds and adults, and the percentages of acceptable responses for these three age groups were 33, 56, and 53 respectively. Twenty-nine percent of the 13-year-olds indicated that they merely translated the given dimension of the square to the "unknown" side, and did not take half of the dimension in order to complete the solution. This was also true for 21 percent of the 17-year-olds and 20 percent of the adults.

Even simple problem-solving sequences are too complex for interpretations to be any more than speculative. One of the major difficulties for 9- and 13-year-olds in the type of problem posed in Exercise RK16 appears to be in carrying out the second stage of what is basically a two-stage problem.

Triangles. Exercise RK08 involved the principle that the sum of the measures of the interior angles of a triangle equals 180 degrees. While the principle is studied as a theorem in most high school geometry courses, it is also dealt with by junior high school students--sometimes through exploratory activities involving measurement or paperfolding. The exercise was administered to each of the three older age groups. The performance level for adults was considerably lower than that for 17-year-olds, perhaps indicating that adults do not find much opportunity for maintaining skill with this principle--assuming the skill was attained in the first place. The results for 13-year-olds would seem to indicate that this age group has not studied the geometric principle involved.

Twenty-six percent of the 13-year-olds indicated that they either added $105° + 50°$ or subtracted $105° - 50°$. Only 10 percent of each of the other two groups appeared to have attempted these two strategies.

Table 5.15
Exercise RK08 and Results

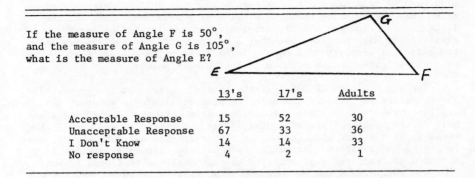

If the measure of Angle F is 50°,
and the measure of Angle G is 105°,
what is the measure of Angle E?

	13's	17's	Adults
Acceptable Response	15	52	30
Unacceptable Response	67	33	36
I Don't Know	14	14	33
No response	4	2	1

<u>Circles</u>. There were six exercises that required some degree of
problem solving related to certain properties of circles. Two of the
exercises were direct applications of the relationship between the
radius and the diameter, d = 2r. Two exercises involved circles with
central angles. Two exercises involved the proportion of the radii
of two circles and their respective areas. A proportion may also have
been used in one of the central angle problems to determine the circum-
ference of the circle; exercises involving proportions are discussed
in another section.

Exercises centering around the relationship d = 2r will be considered first. Exercise RK12 appears to require the straightforward
application of this relationship. The results from exercises RK06 and
RK10 (considered earlier) have indicated that 9-year-olds do not know

Table 5.16
Exercise RK12 and Results

What is the diameter of a circle with a radius of four inches?

	9's	13's
Acceptable Response	5	36
Unacceptable Response	15	31
I Don't Know	75	27
No response	6	6

what the term diameter means, and only 68 percent of the 13-year-olds
correctly identified the diameter of a circle from a picture. Thus,
it comes as no surprise that 75 percent of the 9-year-olds should indicate that they do not know how to attempt Exercise RK12. It is disappointing, however, that only 36 percent of the 13-year-olds were
successful with this exercise. The combined results of Exercises RK06
and RK12 indicate that apparently very little about circles is studied
in grades 4 through 8. The relationship between the radius and the
diameter of a circle is of basic importance to much subsequent mathematics, and certainly should be mastered by a greater proportion of
students prior to high school.

A related unreleased exercise, K30004, also required knowledge of
the relationship $d = 2r$ for its solution. This exercise used a drawing
of a circle that showed various chords (including a diameter) and the
radius. Fifty-one percent of the 13-year-olds and 74 percent of the 17-
year-olds performed the required task successfully. The fact that 13-
year-olds scored a little higher on Exercise K30004 than on RK12 (51
percent compared with 36 percent) might be explained by the fact that
the terms diameter and radius were used in RK12 and no drawing was
present as a visual aid.

Exercise RK19 required knowledge of the relationship between the
radii of two circles and their respective areas.

Table 5.17
Exercise RK19 and Results

What fractional part of the large circle is shaded?			
	9's	13's	17's
Acceptable Response	10	58	70
Unacceptable Response	51	33	19
I Don't Know	35	8	9
No response	4	2	2

It is very difficult to determine what some of the more popular
approaches to Exercise RK19 might have been. While one solution makes
use of the fact that the radius of the shaded circle is half the radius
of the larger circle and the ratio of the areas is therefore 1 to 4, it
is not likely that a large proportion of 13-year-olds used this approach.
In fact, only 20 percent of the 17-year-olds gave acceptable responses
on Exercise K52005 where they were to indicate an understanding of the

proportion involving radii and areas of two circles. About 51 percent gave 2, or 2 times as the relationship between the areas when the radii are in the ratio 2 to 1. This performance by 17-year-olds indicates that 13-year-olds would be even less likely to possess a working knowledge of the proportion involving radii and areas and would not have applied it on Exercise RK19.

It may be that many of the respondents at each age level simply perceived the amount shaded to be one-fourth of the large circle, especially since one-fourth is a familiar quantity to elementary school students.

It is encouraging to note that 70 percent of the 17-year-olds were successful. This is a satisfactory level of performance even if the respondents were estimating rather than using knowledge of the given relationship to solve the problem.

Two unreleased exercises each involved a drawing of a circle with a central angle and its measure clearly labeled. In K30006 the 9- and 13-year-old respondents were to determine the amount, or "fractional part", presumably using the size of the central angle as given. Only 3 percent of the 9-year-olds gave an acceptable response as compared to 28 percent of the 13-year-olds. The exercise could easily have been solved without using the measure of the central angle, or even thinking about the pie-shaped sector as an angle. The circle was marked with quadrant lines; giving the appearance of the usual "fraction-pie" cut into pieces as is found in most elementary school mathematics texts.

Solid Figures

There were two unreleased and one released exercises involving solids that required higher level cognitive abilities than just recognizing or naming figures. Exercise RK07 was administered to the three older age groups. Respondents were to choose the shape with the greatest volume for given dimensions. The exercise is fairly complicated and it is difficult to draw many inferences from it. On exercises such as K20037 and K30027, the data seem to indicate that around 50 percent of the 13-year-olds are familiar with the names and shapes of the cube, sphere, cylinder, and pyramid; although results of this exercise indicate that the 13-year-olds appear to be guessing between the cube, cylinder and pyramid as to the correct shape. It is difficult to determine whether the respondent had in mind any connection between maximum volume and "the most chocolate for his money." The majority of the unsuccessful respondents in the 17-year-old and adult age groups opted for the cylinder or the pyramid. Perhaps each of these two shapes gives an illusion of height, whereas the cube and sphere are perceived to be more compact and thus contain less chocolate. An added complication is that the problem does not state the four chocolates cost the same amount.

Table 5.18
Exercise RK07 and Results

Robert must choose one of four solid chocolate candies to buy.
Which one of the following shapes will give him the MOST
chocolate for his money?

	13's	17's	Adults
Cube one inch on a side	29	50	48
Sphere one inch in diameter	7	6	4
Cylinder one inch in height and one inch in diameter	27	23	20
Pyramid one inch in height with a one inch square base	29	15	13
I Don't Know	8	7	14
No response	1	--	--

There is an interesting connection between two unreleased exercises
involving transformations between a plane figure and a space figure.
In Exercise K20014, 9-year-olds were to select the name of the shape
that would result in cutting open a solid and laying it flat. Thirty-
two percent were successful in this task. In Exercise K20047, the
respondents were to identify the solid shape (from a supplied set of
plastic models) they visualized while watching the exercise administrator spin a plane figure about one axis. Twenty-five percent of the 9-
year-olds were successful as were 47 percent, 61 percent, and 50 percent
of the three older groups respectively.

Little attention is given to transformations or relationships
between plane and solid figures in the mathematics curriculum. It would
seem that more appropriate instructional activities are needed to promote
skill in visualizing relationships between plane and space figures.

Geometric Constructions

Compass and straightedge constructions are studied to some extent
in the seventh through ninth grades, but primarily in plane geometry
classes. The level of performance by 17-year-olds on construction
exercises ranges from 30 to 50 percent correct, and is probably satisfactory given that not all 17-year-olds have taken a course in plane
geometry. A typical exercise is K60001, in which 13- and 17-year-olds
were asked to construct the bisector of a given angle. Thirty-seven
percent of the 17-year-olds were successful on this task, as compared
to only 10 percent of the 13-year-olds. Fifty-five percent of the 13-
year-olds responded "I don't know."

It is unfortunate that more attention is not given to ruler and compass construction in the early secondary school. Many geometric concepts can be more meaningful when applied in construction situations. Certain constructions and general construction procedures are useful in various "real life" applications.

Pythagorean Theorem and Proportions

There were three exercises in which a proportion could be used to solve the problem and three exercises in which the Pythagorean theorem could be used. The former, all unreleased, were administered only to 17-year-olds and the latter were administered to 17-year-olds and adults. These exercises are clustered here for discussion purposes since they represent higher level problem-solving skills.

The three proportion exercises are described and the results given in Table 5.19. Attempts at setting up the proportion were scored as acceptable responses, although this was a negligible percent, usually less than one percent per exercise.

Table 5.19
Results for Proportion Exercises

Exercise Description	Percent of Acceptable Responses for 17-year olds
Determine the circumference of a circle given a central angle and subtended arc length. (K50002)	29
Apply the proportion relation to the ratio of the radii of two circles to the ratio of their areas. (K52005)	20
Find the corresponding side of similar triangles. (K40003)	10

The two released exercises involving the Pythagorean theorem in their solution are shown in Tables 5.20 and 5.21.

Table 5.21
Exercise RK14 and Results

If the length of one leg of a right triangle is nine inches,
and the length of the other leg is 12 inches, how many inches
long is the hypotenuse?

	17's	Adults
Acceptable Response	26	19
Unacceptable Response	51	39
I Don't Know	23	41
No response	1	1

Table 5.21
Exercise RK15 and Results

To set up a tent having the dimensions
shown in the drawing, the vertical tent
poles used should be how many feet high?

	17's	Adults
Acceptable Response	34	29
Unacceptable Response	48	52
I Don't Know	15	19
No response	3	0

It is not profitable to discuss in detail the incorrect responses
to Exercises RK14 and RK15. There was a negligible percent who set up
the problems correctly but made some computational error. The majority
in each case however, either answered "I don't know," or made an attempt
at adding or subtracting the numbers given in the problem. The differences in performance between the two exercises shown above may be
partially explained by the fact that there is some technical vocabulary
in RK14 that would tend to lower performance, whereas RK15 contains a
labeled diagram that would tend to raise performance levels.

These results are similar to those found on unreleased exercise
K51024. This exercise was a "story" problem which required first a
translation of the data to a visualization or diagram of a right triangle, and then application of the Pythagorean theorem to obtain the

solution. Twenty-one percent of the 17-year-olds and 19 percent of
the adults were successful on this exercise.

It is difficult to judge these results as unsatisfactory. The
extent to which 17-year-olds are familiar with these skills is open to
question. Many have probably had some experience with geometric proportions and the Pythagorean theorem, but it is doubtful that the experiences were fresh in their minds or that they were prepared to do much
more than perform some basic manipulation with these skills. It is not
the case that one is to be content with a range of 10 to 34 percent
success with these exercises, but it is difficult to foresee what
strengthening the curriculum in such areas would produce in the way of
higher performance.

Summary

Performance of the 9-year-olds was typically rather poor on exercises that required them to apply fundamental relationships between
plane figures. Only a few could find the diameter of a circle given its
radius (5 percent), find the fraction of a circle determined by a given
central angle (3 percent), or find the distance between the centers of
two adjacent squares (36 percent).

Thirteen-year-olds uniformly performed at a substantially higher
level than 9-year-olds (usually 25 to 35 percent higher), although their
performance was moderately low on topics such as the diameter-radius
relationship (36 percent acceptable responses) and finding the fraction
of a circle determined by a given central angle (28 percent). Performance was also low on exercises involving angle measurement and relationships: fewer than 40 percent measured angles acceptably with a
protractor; only 10 percent were able to construct an angle bisector;
85 percent were unable to use successfully the principle on the sum of
the interior angles of a triangle.

The performance level of the 17-year-olds was substantially higher
than that of the 13-year-olds, although their level of performance was
not uniformly acceptable, and was particularly low on measuring and
bisecting angles. The percent of acceptable responses for exercises
involving the application of a proportion or the Pythagorean theorem
fell in a range of from 10 to 34 percent for 17-year-olds.

In general, adults did not perform as well as the 17-year-olds,
but performed better than 13-year-olds on most geometry exercises.

Implications for Instruction

In most elementary schools geometry should have a more prominent place in the curriculum, and at every grade appropriate attention should be given to the introduction, development, and maintenance of geometric concepts. The role of geometry in the elementary school is often debated. Geometry topics frequently receive a low priority in mathematics instruction. Many mathematics educators recommend the inclusion of selected geometry topics for enrichment purposes. Some feel that geometric concepts can help later in developing computational skills and problem-solving abilities.

Much geometry instruction in the primary grades can be devoted to exploratory and introductory activities in which pupils can see, touch, and manipulate models of solid as well as plane figures. Although teachers may not expect mastery of such technical terms as sphere, parallel, diameter, etc., these terms should be used consistently. The student should learn that although "ball" refers to an example of a sphere, it is not the proper name.

Pictures, models, and everyday examples of solid figures should be carefully related to each other using the technical terms. Some relationships between plane and solid figures should be explored. An orange can be used to illustrate the sphere and then sliced to show the circle. A cardboard triangle can be rotated rapidly about an altitude to generate a cone. Pictures illustrating real-life examples of parallel lines, angles, or diameters of circles can be displayed on the bulletin board. Time for activities like these might be found by reducing somewhat the time currently devoted to learning to recognize such common plane shapes as the square, circle, and triangle.

More and more adequate instructional materials are needed in recognizing and applying relationships involving circles, certain polygons, parallel lines, and angles. This should include some introductory work with the Pythagorean theorem and geometric proportions, but at an intuitive level. Successful performance in applying geometric relationships depends, to a large extent, upon a sound initial development of the basic concepts and terminology. The responsibility for developing a high level of performance with many of the key geometric relationships contained in the assessment exercises should fall on the upper elementary through junior high school grades. The performance by the 13-year-olds on exercises involving angles was particularly disappointing. Perhaps some laboratory activities on angle measurement and angle construction, including bisection, would strengthen pupils' understanding of the angle concept.

The technical vocabulary introduced in the elementary grades should be mastered and expanded during the early secondary grades. This should include an emphasis on being able to name common solid figures, use technical terminology, and explain relationships between plane and solid figures.

The assessment did not adequately sample the diversity of geometry studied in secondary schools.   For example there were no exercises on congruence, proof, transformational geometry, or non-Euclidean geometry. The assessment results imply, however, that more emphasis is needed on learning Euclidean principles and relationships, and applying them to problem situations.

VI

Measurement

Forty-five exercises were administered that tested measurement concepts. Twenty-six exercises were released, and seven were individually administered. The results do not provide a comprehensive view of measurement since a number of basic concepts were only measured tangentially or not at all. In addition, it is clear that the context of the problem and the numbers involved in the exercise significantly affected performance. Therefore, one must exercise some caution in drawing general conclusions about the development of measurement concepts.

For the purpose of discussion, exercises have been classified in the following general categories: 1) Converting and comparing units of measure; 2) Estimating and measuring length; 3) Perimeter, area, and volume; 4) Maps and scale drawing; 5) Time; and 6) Using measuring instruments.

Converting and Comparing Units of Measure

Overview of Results

Thirteen exercises (six released) were administered that required students to convert from one unit of measure to another. These included all of the problems involving weight (three exercises) and liquid capacity (four exercises) as well as three exercises involving measuring length and three exercises involving metric measurement. A fourteenth exercise (unreleased) required subjects to compare quantities given in different units of measure (e.g., 11 inches to 1 foot). Most of the conversion exercises employed problem situations, and the context of the problem frequently affected the difficulty of the exercise. Therefore, the data presented generally represent a measure of more than just the ability to convert units.

Following are the major observations from the exercises on converting and comparing units of measure:

I. All age groups are successfully able to compare quantities measured in pints or quarts and feet or yards. Thirteen-year-olds, 17-year-olds, and adults are also able to successfully compare ounces to pounds and weeks to months, but 9-year-olds were generally unsuccessful in making these comparisons.

II. Converting from one unit of measure to another was a difficult task for 9-and 13-year-olds, especially where fractional units were involved. Most 17-year-olds and adults could make such conversions in a simple problem situation, but only an average of one-third of both groups could successfully convert from one unit of measure to another in a more complex problem situation.

III. Most 13-year-olds were unfamiliar with metric units.

IV. The metric system has not become a part of everyday experience for most adults, as fewer than half were familiar with standard metric units of measure, a drop of 10 percent from the performance of 17-year-olds.

V. Thirteen-year-olds showed substantial gains over the performance of 9-year-olds on every exercise administered to both groups. Similarly, the performance of the 17-year-olds over the 13-year-olds was a substantial improvement. Adults performed at about the same level as the 17-year-olds.

Comparing Units of Measure

The results summarized in Table 6.1 indicate that most 13-year-olds, 17-year-olds, and adults successfully compared quantities measured with different units of measure (e.g., 11 inches to 1 foot, 5000 pounds to 2 tons). Nine-year-olds were relatively successful with comparisons involving yards and feet and pints and quarts but had significantly greater difficulty with comparisons involving ounces and pounds and months and weeks. Across all age levels the comparison involving ounces and pounds was the most difficult.

Table 6.1

Comparison of Quantities Given in
Different Units of Measure (Unreleased Item)

Units	9's	13's	17's	Adults
Feet-Yards	82	94	97	96
Ounces-Pounds	36	73	85	92
Pints-Quarts	83	92	94	95
Weeks-Months	54	87	94	99

Approximately 36 percent of the 9-year-olds, 73 percent of the 13-year-olds, 85 percent of the 17-year-olds, and 92 percent of the adults successfully compared quantities measured in ounces and pounds. Comparisons involving liquid capacity were somewhat easier, especially for the younger respondents, as over 83 percent of the 9-year-olds and over 92 percent of the other populations successfully compared quantitites measured with pints and quarts.

However, on another exercise (RE15), only 44 percent of the 9-year-olds and 84 percent of the 13-year-olds could identify the number of quarts in a gallon.

Approximately 82 percent of the 9-year-olds and 95 percent of the other populations successfully compared lengths given in feet and in yards (see Table 6.1). However, only about 45 percent of the 13-year-olds and 65 percent of the 17-year-olds and adults successfully solved problems involving conversion of measures given in feet and inches to inches and feet to yards.

The results summarized in Tables 6.2 and 6.3 indicate that the ability to apply knowledge of unit conversion was substantially influenced by the complexity of the problem situation. Performance on exercise RE11 was within six percentage points of performance on the items involving a comparison of ounces and pounds, whereas performance on exercise RE13 was substantially below (over 50 percent) performance on either the item involving a comparison of pints and quarts or the item involving a comparison of quarts and gallons. Data from unreleased exercises also indicate that problems involving fractional units (e.g., 1/2 pound, 3 1/2 quarts) were significantly more difficult at every age.

Table 6.2
Exercise RE11 and Results

A man bought two pounds of cheese in eight-ounce packages. How many packages did he buy?

	17's	Adults
Acceptable Response	81	86
Unacceptable Response	14	11
I Don't Know	5	3
No Response	1	0

Table 6.3
Exercise RE13 and Results

A recipe for punch calls for equal amounts of lemonade, limeade, orange juice and ginger ale. How many PINTS of ginger ale would be needed in order to make two gallons of this punch?

	13's	17's	Adults
Acceptable Response	17	30	38
Unacceptable Response	68	59	55
I Don't Know	13	9	8
No Response	2	1	0

Metric Units

Two unreleased exercises tested recognition of metric units of measure. In one exercise respondents were asked to identify the largest metric unit of measure from a list of four, and in the other respondents were asked to identify which English unit approximately corresponded to a given metric unit. About 56 percent of the 17-year-olds and 45 percent of the adults correctly answered each question, and about 37 percent of the 13-year-olds identified the largest unit.

Few subjects among the 13-or 17-year-olds were able to use the conversion formula to change a Fahrenheit temperature to Centigrade (see Table 6.4). However, this result reflects algebraic abilities more than knowledge of the metric system.

Table 6.4
Exercise RH07 and Results

The formula for the relationship between Fahrenheit and Centigrade temperatures is $F = \frac{9}{5} C + 32°$. What is C when F is $77°$?

	13's	17's
Acceptable Response	2	29
Unacceptable Response	43	33
I Don't Know	47	35
No Response	8	3

Summary

All age groups except 9-year-olds successfully compared quantities measured with different units of measure, and 9-year-olds were relatively successful with comparisons involving yards and feet and pints and quarts. However, specific conversions were significantly more difficult for all age levels. With increased use of the metric system, which does not require complex conversion factors, this type of problem will be less frequently encountered in everyday experience. Since pupils are generally familiar with the relations between units, the time and effort required to overcome difficulties converting units does not appear to be warranted. Instead, teachers should focus on the more basic measurement concepts discussed below. However, the fact that just 45 percent of the adults and 56 percent of the 17-year-olds demonstrated any familiarity with metric units indicates that the metric system is not yet a part of the everyday experience of most of the population.

Estimating and Measuring Length

Overview of Results

In addition to the three exercises discussed in the previous section dealing with converting units of length, three were administered that required respondents to measure length and one that required them to estimate the length of a line. In the estimation problem (unreleased) 39 percent of the 9-year-olds correctly estimated the length of a line in inches.

The three exercises testing the ability to use a ruler and measure length were individually administered. In one exercise, 9-and 13-year-olds were asked to identify four points on a ruler. They were also asked to use the ruler to measure the length and width of a rectangular board and to measure a curve using a trundle wheel. In another exercise, all age groups were asked to determine the thickness of the bottom of a box (1 inch), which required finding the difference between the inside and outside depth. In the third exercise, all four age groups were asked to find the "distance around" a cylinder using a ruler and piece of string.

The following major observations may be made from the data on estimating and measuring length.

I.   Most 9-and 13-year-olds are able to measure lengths shorter than a ruler where the measure is a whole number of inches. However, fewer than half of the 9-year-olds are able to measure lengths longer than a ruler.

II.   Measuring length in fractional units is a difficult task for 9-and 13-year-olds.

III. Making an indirect measurement of length is a more difficult task than making a direct measurement.

Direct and Indirect Measurement

The results summarized in Table 6.5 indicate that although most respondents could identify points on a ruler that represented a whole number of inches, substantial difficulty was encountered with fractional parts of an inch, especially with fractions other than 1/2. In fact, 29 percent of the 9-year-olds and 13 percent of the 13-year-olds identified 3 3/8 inches as 3 1/2 inches. Most respondents were able to measure lengths less than 12 inches (see Table 6.6) but had significant difficulty measuring lengths greater than 12 inches, where it was necessary to move the ruler and find the sum of the measures.

Not surprisingly, respondents had greater difficulty measuring with a trundle wheel than with a ruler (see Table 6.7). It is interesting to note that most respondents indicated that

Table 6.5
Reading a Ruler

Here is a ruler with four points: A, B, C, and D marked on it.
What is the reading at point _____?

		9's	13's
A:	3/4 inch	14	54
B.	4 1/2 inches	60	83
C.	2 inches	84	93
D.	3 3/8 inches	2	25

Table 6.6
Using a Ruler

Use the ruler to measure the length and width of this board. What
is the width of this board? (Point to width of board). What is
the length of this board? (Point to length of board).

	9's	13's
Width: 7 inches	82	91
Length: 15 inches	48	73

Table 6.7
Using a Trundle Wheel

A. (Pick up trundle wheel.) Do you think you can use this instru-
    ment to measure the length of the line on this page? (Point to
    line.)

B. Use this instrument to measure the line on this page. (Point
    to line.) What is the length of the line?

	9's	13's
A: Yes	68	81
B: Correct - 8 1/2 inches	16	39
± 1/8 inch		

---

they could use the trundle wheel to measure the length of the
curve, but few of them were actually able to do so. A common
error that one might expect children to make with a trundle wheel
is to ignore complete revolutions of the wheel. In this problem
in which a complete revolution of the trundle wheel indicates a
measure of five inches and the curve is eight inches long, this
error would yield an answer of three inches. However, this error
accounted for fewer than 3 percent of the 9-year-olds' responses
and fewer than 1 percent of the 13-year-olds' responses.

Measurement problems requiring indirect measurement were more
difficult than direct measurement exercises (see Table 6.8) with
the decrease in performance substantially greater for 9-year-olds
than for 13-year-olds.

Table 6.8
Indirect Measurement of Length

A. Use the ruler to determine the thickness of the bottom of this
box.

B. Here are a cylinder, ruler, and piece of string. What is the
distance around the cylinder?

	9's	13's	17's	Adults
A.	18	43	60	64
B.	23	59	70	82

Summary

About 83 percent of the 9-year-olds and 93 percent of the 13-
year-olds could successfully operate with the simplest measurement
processes--comparing units, identifying a whole number of units, and
measuring lengths shorter than the ruler. Any complexity (fractional
units, indirect measurement, etc.) had a much greater effect on the
performance of 9-year-olds than 13-year-olds. In fact, fewer than
50 percent of the 9-year-olds could deal with any of these complex-
ities, with the exception of identifying half units.

Perimeter, Area, and Volume

Overview of Results

Ten items (six released) dealt with perimeter, area, and volume.
Although the ten items do not provide a comprehensive sampling of
perimeter, area, and volume concepts, the results for virtually every
item administered indicate that at all levels respondents encountered

significant difficulty with these concepts. In only two of the 20 response categories were more than 50 percent of the responses correct.

The following major observations may be made:

  I.   Nine-and 13-year-olds appear to have little knowledge of basic area concepts, but most 17-year-olds and adults were able to calculate area in a simple problem situation.

  II. The concept of perimeter is evidently not well established in 9-year-olds.

  III. The concepts of volume and surface area are easily confused by 13-and 17-year-olds.

Concepts of Area and Volume

In exercise RK17 (see Table 6.9) only 38 percent of the 9-year-olds demonstrated any ability to deal with a basic area concept comparing the number of square units covering each figure. Forty-four percent of the respondents chose the 3 x 5 rectangle as having the same area as the 4 x 4 square. One of several explanations may account for this response. Respondents have ignored the number of units and chosen the figure whose shape was most similar to the standard figure. This tendency to judge area strictly on the basis of physical appearance has been identified in various Piagetian studies. A second plausible explanation for this error is that respondents compared the sums of the length and width of each figure, thus confusing area with an incorrect notion of perimeter.

Table 6.9
Exercise RK17 and Results

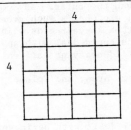

Which of the figures below has the same area as the figure above?

Table 6.9 continued

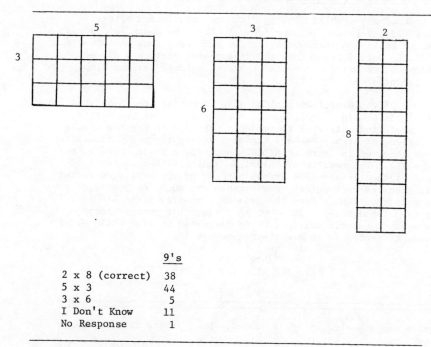

	9's
2 x 8 (correct)	38
5 x 3	44
3 x 6	5
I Don't Know	11
No Response	1

In a similar volume exercise (unreleased 9-, 13-, and 17-year olds were shown a picture of a rectangular solid cut into cubes and asked to find the number of cubes contained in the solid. This exercise is considerably more difficult than the area problem, as it is not possible to simply count the units of measure. One must recognize that certain cubes are not pictured and must multiply or in some other way compute the number of cubes. Six, 21 and 43 percent respectively of the 9-, 13-, and 17-year-olds correctly answered this question. There was a marked tendency, which decreased significantly with age, to simply count the number of unit squares on the three faces pictured. Forty-eight percent, 27 percent, and 13 percent of the respondents chose this type of response. Older respondents recognized that certain cubes were not pictured; but this led many to indicate that the volume was twice the number of squares pictured on the three faces, thus accounting for the other three faces but confusing a surface area notion with volume. This response accounted for 2 percent, 13 percent, and 13 percent respectively.

Computing Area and Perimeter

In the only problem dealing with the calculation of perimeter (PK13), 9-year-olds were asked to find the amount of fencing
needed to enclose a rectangular garden nine feet long and five
feet wide. Only 7 percent of the 9-year-olds correctly solved
the problem; 8 percent computed the area, and 43 percent simply
added the nine and five.

None of the exercises dealing with computing area consisted
of straight-forward calculations; consequently, performance is
not entirely indicative of respondents' ability to perform simple
calculations of area. Perhaps the most straightforward calculation
of area occurred in Exercise EI3002 involving a scale drawing (see
Table 6.11). The results of this item indicate that at least 58
percent of the 17-year-olds and 69 percent of the adults could calculate areas of rectangles. Performance was much lower when respondents were required to find the area of a square, given its perimeter (see Table 6.10). It is worth noting that younger respondents most frequently multiplied the perimeter by four while older
respondents tended to square the perimeter.

Table 6.10
Exercise RE10 and Results

A square has a perimeter of 12 inches. What is its AREA in square
inches?

	13's	17's	Adults
Acceptable Response	7	28	27
48	20	10	10
144	12	19	25
Other Unacceptable	33	23	18
I Don't Know and No Response	28	19	21

Calculating the area of a rectangle with an interior rectangle removed presented similar difficulties, as less than a third
of the 17-year-olds and about 40 percent of the adults could find the
area of the resulting region.

Problem Situations

Forty-six percent of the 17-year-olds and 63 percent of the
adults were able to calculate the quantity of asphalt paint necessary to cover a rectangular driveway. However, fewer than 28 percent of the 17-year-olds and 39 percent of the adults could successfully solve problems which required converting square feet to square
yards. In these problems 43 percent of the 17-year-olds and 29 percent of the adults multiplied the number of square feet by one-third.

Summary

At all levels for virtually every exercise, perimeter, area, and volume problems were exceptionally difficult. Over half of the 9-year-olds demonstrated little comprehension of the basic area concept of covering with unit squares, and similar deficiencies were shown for volume by 13-and 17-year-olds. Problems requiring area calculations were even more difficult, except in the simplest cases involving areas of rectangles.

Maps and Scale Drawing

Overview of Results

Four exercises were administered that dealt with finding distance or dimensions on maps and scale drawings. The following observations may be made from the data gathered on these exercises:

I.   Most 17-year-olds and adults can successfully find dimensions on a scale drawing and distances on a map when the measurement involved is a whole number of inches.

II. Few 17-year-olds and adults were able to find the dimensions of an irregularly shaped room from a scale drawing and subsequently calculate the area of the room.

In an individually administered exercise, 17-year-olds and adults were given a scale drawing (see Table 6.11) and a ruler and asked to find the dimensions of the rooms. Over 85 percent of both age groups could successfully find dimensions when the measurement involved a whole number of inches. Measurements involving a fraction of an inch were significantly more difficult. Approximately the same percent of 17-year-olds and adults successfully performed similar calculations involving finding distances on maps. These map problems were also administered to 9-and 13-year-olds with 40 percent and 74 percent respectively solving them.

In measuring Room 2 of the scale drawing (which was not a rectangle) over 16 percent of the respondents failed to attempt all the measurements of the room, in spite of the fact that there were arrows on the diagram to indicate which dimensions were to be measured. This omission may have resulted from an inference of the respondents that Room 2 was rectangular-shaped and did not include the vestibule area; or it may have resulted from an inability to identify the relevant dimensions of irregular shapes, which would indicate a serious deficiency.

Table 6.11
Exercise E13002 and Results

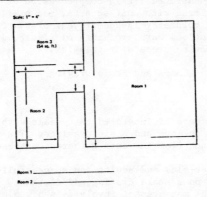

1. This is a scale drawing of the rooms of a house (point to the scale drawing). One inch on the drawing represents four feet. Use the ruler to measure the length of the walls in Room One. Write your measurements in feet between the appropriate arrows. (Point to the set of arrows to show respondent where to write his answer.)

2. Now measure the length of the walls in Room Two. Write your answer like you did for the first room.

			17's	Adults
Room 1:	horizontal - 16 feet		85	87
	vertical - - 18 feet		70	78
Room 2:	Bottom	6 feet	72	76
	left	12 feet	83	85
	top	10 feet	66	73
	right	4 feet	73	71
Room 2:	No Response			
	bottom		3	6
	left		2	3
	top		12	11
	right		16	20
Area:	Room 1		58	69
	Room 2		29	40

Time

Overview of Results

Eleven exercises (six of which are released) were administered that dealt with time concepts. These exercises can be classified into three groups of related items: 1) exercises dealing with reading clocks; 2) exercises dealing with finding the time (or date) a given number hours (or days) before or after a given time (or date); and 3) exercises dealing with finding the length of time between two given hours.

The following general observations may be made:

I.   Most 9-year-olds appear to be able to read a clock and tell time.

II. The ability to determine the amount of time between two given hours appears to be relatively easy for a majority of 13-year-olds, but a fairly difficult task for 9-year-olds.

III. Determining the amount of time between two given times in a complex problem situation appears to be difficult for most 17-year-olds and adults.

Reading Clocks

The only exercise dealing with reading clocks was an individually administered unreleased exercise. The results for this exercise indicate that 96 percent of 9-year-olds can correctly tell time on the hour (e.g., 5:00 o'clock), 80 percent can correctly tell time on the half hour (e.g., 2:30), and 73 percent can correctly tell time with five minute intervals (e.g., 7:15, 6:05).

Operations with Time

The results summarized in Table 6.12 indicate that approximately 65 percent of the 13-year-olds and 84 percent of the 17-year-olds and adults successfully solved exercise RE03, in which they were required to find the time 10 hours after 7:45. Just over 40 percent of the 9-year-olds successfully solved a similar non-released exercise.

Problems involving finding the length of time between two given hours proved to be uniformly more difficult. Only 25 percent of the 9-year-olds successfully computed the number of minutes between 4:25 p.m. and 5:00 p.m.

Table 6.12
Exercise RE03 and Results

A worker went to his job at 7:45 a.m. and returned home exactly 10
hours later. At what time did he reach home?

	13's	17's	Adults
Acceptable Response	65	84	84
Unacceptable Response	31	15	15
I Don't Know	3	1	1
No Response	1		

Using Measuring Instruments

Overview of Results

    In addition to measuring length (discussed in a previous
section), respondents were asked to read scales and perform
measurement operations involving angles, temperature, and weight.

    The following general observations may be made:

    I.   About one-third of the 13-year-olds and half
         of the 17-year-olds and adults could use a
         protractor to measure angles.

    II. Over ninety percent of the 9-and 13-year-olds
         could read temperatures on a thermometer
         that were labeled on the scale, but fewer
         than a quarter of the 9-year-olds could
         read temperatures that were not labeled or
         were below zero. More than half of the 13-
         year-olds could perform this task.

    About a third of the 13-year-olds and half of the 17-year-
olds and adults could use a protractor to measure angles. On the
whole, there was no great difference in difficulty between measuring angles that were a multiple of 10 and those that were not,
or between measuring angles less than 90 and those greater.

    Younger respondents, on the other hand, had significantly
greater difficulty in reading temperatures on a thermometer when
they were not labeled on the scale of the thermometer. The
difficulty was caused by the fact that only the even degrees were
marked on the scale and only multiples of ten were labeled. Although over 90 percent of 9-and 13-year-olds could identify temperatures that were a multiple of ten, only 19 percent of the

9-year-olds and 56 percent of the 13-year-olds could identify a temperature that was not. Fifty percent of the 9-year-olds and 36 percent of the 13-year-olds incorrectly assumed that each mark represented an interval of one degree. Both groups also had substantial difficulty identifying temperatures below zero. Only 23 percent of the 9-year-olds and 52 percent of the 13-year-olds correctly identified a below zero temperature. Since thermometers are often used to motivate negative integers, it is of some significance that, for close to half of the 13-years, below zero temperatures held little meaning.

Implications for Instruction

Pupils need experience with a wide variety of measurement situations. Perhaps it is easier to have pupils measure distances printed on a piece of paper, but the results indicate that measuring distances longer than the ruler is a skill that does not generalize from these experiences. Pupils need practice measuring longer distances. The poor performance on exercises requiring calculating the area from the perimeter indicates that many pupils apply formulas rotely. Practice is needed with two-step problems, in which students must decide what dimensions are needed to calculate the area or perimeter, and therefore, cannot rotely plug numbers into formulas. Difficulties with fractional units also indicate that practice is needed in measuring with fractions other than 1/2.

Substantial hands-on experience with physical models of geometric shapes is needed. Pupils are not intuitively familiar with concepts of perimeter, area, and volume. They need experience in partitioning regions into unit squares and solids into unit cubes and counting the units. They need experience measuring the distance around geometric shapes. Perimeter, area, and volume formulas should not be derived until pupils are confident in working with the physical models and have some insight into how the formula is derived. Furthermore, when in doubt, pupils should be encouraged to return to pictures or other physical models for verification.

Specific attention should be devoted to the concept of a unit of measure. Much of the difficulty in the area and volume exercises resulted from pupils ignoring the units of measure or incorrectly identifying the units of measure. Incorrectly identifying the unit also accounted for the difficulties in reading a theromometer.

The process of measurement is one of the primary ways that abstract mathematics is applied to the physical world. The need for measurement operations occurs naturally in everyday experience. Furthermore, measurement concepts reinforce basic number concepts. However, measurement concepts do not develop naturally without experience, and even with experience they require substantial preparation and background. Therefore, above all, teachers are encouraged to provide an abundance of measurement experiences in every grade.

VII
Consumer Mathematics

Consumer mathematics represented a diverse area and included a wide range of exercises. In an effort to organize these exercises and to simplify their presentation, a classification into three groups was made: 1) Average; 2) Graphing; and 3) Purchase and Cost. This represents one classification scheme, but it was clearly not the only one that could have been generated (see Figure 1). Each of the classifications is accompanied by a brief description used to place the consumer exercises into appropriate groups or clusters. Figure 1 also reports the age groups assessed with each released exercise. It would be expected that relevant consumer problems are more frequently encountered by the two older age groups and this is reflected by the fact that more than 90 percent of the consumer mathematics exercises were given to 17-year-olds and/or adults. Less than half of the consumer mathematics exercises were given to 13-year-olds and only one exercise was given to 9-year-olds. Approximately the same proportion of unreleased consumer mathematics exercises was administered to each of the respective age groups.

Classification	Release Number of Exercise	Age Groups Sampled by Exercises	Description of Exercises
Average	RC12	17 A	Exercises involving the
	RC13	9 13 17 A	calculation of averages or
	RJ06	13 17 A	requiring the use of averages in their solution.
Graphing	RH09	13 17 A	Exercises that require con-
	RQ02	17 A	structing graphs or reading
	RQ03	13	and/or interpreting data
	RQ04	17 A	in either graphical or
	RQ05	13 17 A	tabular form
	RQ06	17 A	
Purchase and Cost	RP01	13 17 A	Exercises presenting typical
	RP02	17 A	consumer problems encounter-
	RP03	17 A	ed by the citizen. All of
	RP04	A	these problems require the
	RP05	17 A	determination of costs or
	RP06	A	expenses.
	RP07	13 17 A	
	RC26	17 A	

Figure 1. Classification and Description of Clusters of Consumer Mathematics Exercises

Discussion of these groups of exercises will be done in the order shown in Figure 1. The presentation will concentrate on released exercises. Selected unreleased exercises will also be discussed whenever these findings conflict with results from released exercises or when they provide specific information that is unavailable from the released pool of exercises. A case in point for this latter situation involves several individually administered exercises. Discussion of these exercises or portions of them are integrated as appropriate.

Average

Overview of Results

Exercises discussed in this section involved either the calculation of averages or made use of averages in their solution. Following are the major observations that may be made from the results of these exercises.

I.   Levels of performance on exercises involving calculation of arithmetic means were high, ranging from 75 and 90 percent correct for both 17-year-olds and adults.

II. Seventeen-year-olds and adults experienced difficulty in calculating averages other than arithmetic means, with percentages of correct responses ranging from approximately 20 to 50.

III. Between 5 and 10 percent of each age group assessed responded "I don't know" to each exercise that involved finding averages.

Exercises Involving Averages

Exercise RJ06 is a typical problem involving averages. The fact that this was a simple exercise no doubt contributed to the high level of performance (see Table 7.1). Approximately three-fourths of both the 17-year-olds and adults responded correctly to the exercise, although only about half of the 13-year-olds did so. (This pattern of results was consistently found in the consumer exercises with the 17-year-olds and adults receiving approximately the same percentage correct, and the 13-year-olds scoring much lower.) It was surprising to find that 7 percent of the adults responded "I don't know" to this exercise, since averages are widely used and general knowledge of them tacitly assumed.

An unreleased exercise (J21009) required respondents to find the median for a set of data. Less than 20 percent of the 13's, 17's and adults responded correctly.

Table 7.1
Exercise RJ06 and Results

Last summer Todd earned $205, Charlotte earned $562 and Dale earned
$400. What is the average of their summer incomes?

	13's	17's	Adults
Acceptable Responses			
$389, $400	48	74	77
Unacceptable Responses			
$1167 and others	46	21	16
I Don't Know	5	4	7
No response	1	1	0

Although the topic of average is not usually discussed until the
intermediate grades or later, one of the exercises using averages
(RC13) was given to 9-year-olds, as well as the other three age groups.
The respective percentages correct were 6, 44, 68, and 69. The exercise gave an average speed of 50 miles per hour and asked how long it
would take to go 275 miles. There is no reason to expect 9-year-olds
to know the concept of average or to be able to apply the division
operation needed to answer the question.

Summary

Assessment exercises involving averages were generally very simple
and required only direct computation. Performance on determining
arithmetic averages was high (75-90 percent correct) for both 17-year-
olds and adults. Calculation of other averages, such as a weighted
mean and median, were much more difficult, indicating this area deserves
more attention in mathematics programs.

Graphing

Overview of Results

The graphing cluster is characterized by exercises that involve
reading and interpreting either from graphs or tables of data. High
levels of performance were obtained on several exercises within this
group as basic graph reading skills seem to be reasonably well developed
for all age groups. Following are the major observations for which
supporting data will be presented.

I. Direct interpretation of pictorial and bar graphs was
easy (more than 90 percent correct) for 13-year-olds.

II. Marked improvement in graph reading skills was observed
between 13-year-olds and 17-year-olds, with little
difference in performance between 17-year-olds and
adults. A majority of these two age groups correctly
responded to each graph reading exercise.

III. In all age groups, direct interpretations were much
easier than problems requiring not only interpretations
but judgments and decisions.

Selected Graphing Exercises

A careful examination of the graphing exercises suggests that the
context of the question is probably the single most influential factor
on performance levels. A case in point involves exercises RQ03
(Table 7.2) and RQ04 (Table 7.3). Each provides similar information,
but the percentages of correct responses vary greatly. Whereas over
90 percent of the 13-year-olds correctly answered RQ03, only 55 and
62 percent of the 17-year-olds and adults answered RQ04. One exercise
(RQ03) required a direct reading of the graph; the other (RQ04) required
a comparison between two bits of data displayed by the graph. This
additional task of comparing data from two groups and then estimating
a ratio increased the difficulty considerably.

Table 7.2
Exercise RQ03 and Results

		Region	Rural Population (0 = 1 million persons)
	1.	New England	0 0 0
	2.	Middle Atlantic	0 0 0 0 0 0
U. S. Rural	3.	East North Central	0 0 0 0 0 0 0 0 0
Population for	4.	West North Central	0 0 0 0 0
Nine Regions in	5.	South Atlantic	0 0 0 0 0 0 0 0 0 0
1970	6.	East South Central	0 0 0 0 0
	7.	West South Central	0 0 0 0 0
	8.	Mountain	0 0
	9.	Pacific	0 0 0 0

According to the chart, which TWO regions of the U. S. had the largest
rural populations in 1970?

	13's
Acceptable Responses	
3 and 5, 5 and no other answer	91
Unacceptable Responses	6
I Don't Know	2
No response	2

Table 7.3
Exercise RQ04 and Results

Number of Telephones in Operation in Various
World Areas in 1968
(0 = 8 million telephones)

Area	Number of Telephones
North America	0 0 0 0 0 0 0 0 0 0 0 0 0 0 0 0 0
Europe	0 0 0 0 0 0 0 0 0 (
Asia	0 0 0 0

According to the chart, in 1968 the number of telephones in operation in North America was how many times the number of telephones in operation in Asia?

	17's	Adults
Acceptable Response		
4, 1/4 as many as in North America	55	62
Unacceptable Responses		
12, 16, and others	39	30
I Don't Know	5	7
No response	2	1

Another example of the context of an exercise influencing the difficulty is found in RQ05 (Table 7.4). Although the same graphical data were presented, approximately 40 percent less of each age group answered the second part of RQ05 than the first part. The substantial increase in the percent of "I don't know" responses may be that terms such as maximum and minimum in the second part of RQ05 caused confusion although a close examination of the incorrect responses to RQ05B suggests a different explanation. Approximately one-fifth of each age group reported their answers in units (2.8-3.2) rather than thousands (2800-3200) of units. These respondents solved the problem, but used incorrect units to express the answer. Consequently, the performance on this exercise may not be as bleak as Table 7.4 suggests.

Table 7.4
Exercise RQ05 and Results

The graph shows the monthly production of Company X during 1971.

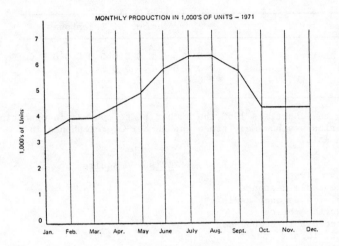

MONTHLY PRODUCTION IN 1,000'S OF UNITS – 1971

A. The greatest drop in production from one month to the next
    occurred between what two months?

Answer _____ to _____

	13's	17's	Adults
Acceptable Responses			
Sept. to Oct., 9 to 10	45	69	63
Unacceptable Responses	49	28	33
I Don't Know	5	2	4
No response	2	1	0

B. The difference between maximum and minimum monthly production
    during 1971 was approximately how many units?

Answer _____

	13's	17's	Adults
Acceptable Responses	7	21	25
Unacceptable Responses	68	64	58
I Don't Know	22	12	16
No response	3	4	1

RQ06 was the only exercise involving only tabular data that was
included in this group. It was found that the performance levels for
17-year-olds and adults on this exercise were within the same range
as was found with exercises involving graphical data, namely 50 to 70
percent correct. Most of the consumer exercises using only tabular
data also involved cost and money. Therefore additional discussion
related to performance levels for reading and interpreting tables is
found in the purchase and cost section.

Summary

Direct interpretation of pictorial and bar graphs was easy (more
than 90 percent correct) for 13-year-olds. With the single exception
of RQ05, a majority (ranging from 50 to 78 percent) of the 17-year-olds
and adults answered each of the exercises in this group. There was
little difference in performance levels between these age groups on any
of these exercises and no consistent difference in performance levels
appeared. Performance levels for each age group were influenced by the
context of the exercise. Direct interpretations from graphs were accompanied by a much higher percent of correct responses than exercises
requiring not only interpretations but judgments and decisions.

Purchase and Cost

Overview of Results

Each purchase and cost exercise required the determination of cost
or expenses in a problem situation. The following major observations
may be made.

  I.   Typical consumer problems that required more than one
       operation were difficult for all respondents

 II.   Skills in reading and interpreting sales tax tables were
       high for both 17-year-olds and adults, but performances
       were much lower on problems that required reading federal
       income tax tables.

III.   Problems that required a determination of the best buy
       from available choices were difficult for a majority of
       respondents.

 IV.   Skills in estimating and computing prices appear to be
       reasonably well developed among the older age groups.

Selected Purchase and Cost Exercises

Exercise RP01 (Table 7.5) presented a realistic consumer problem that was correctly solved by approximately half of the 13-year-olds and by more than three-fourths of the two older age groups. The context of this exercise was simple, the solution required little calculation, and the numerical values used (10%, 15% and $100) helped make this an easy problem. This claim is supported by an examination of the type and frequency of error patterns for RP01 and similar exercises. There were only three frequently occurring incorrect responses, namely 10, 15, and 25. Furthermore, few direct computational errors were detected. Incidentally, a correct answer for this exercise could have been obtained by simply subtracting the smaller percent from the larger and reporting the difference. No data related to the problem-solving process were available, so the extent to which this strategy was used in not known.

Table 7.5
Exercise RP01 and Results

Television sets are on sale at two stores. One offers a 10 percent discount while the other offers 15 percent. What is the difference in the sale price at the two stores of a TV set that is regularly priced at $100?

	13's	17's	Adults
Acceptable Responses			
$5, 5	49	76	86
Unacceptable Responses			
$25, 15, 10	28	15	9
I Don't Know	22	8	5
No response	2	1	0

Compare the context of RP01 with that of exercise RC26 as reported in Table 7.6. RC26 presented a problem that required the respondent to determine the expenses for parking a car in a lot. The information presented in the problem, relating total cost to time and money, is more complex than that of exercise RP01. Although an algorithm can be used to determine the parking cost, the particular function must be developed from careful translation of the information provided in the exercise. Then the appropriate arithmetic data from the exercise must be substituted before any calculation can be made. No doubt each of these factors (development of an algorithm, proper substitution and calculation) took its toll of respondents; consequently, each of these factors should be given careful attention by teachers when interpreting test data, writing consumer problems or merely assigning such problems.

Table 7.6
Exercise RC26 and Results

A parking lot charges 35 cents for the first hour and 25 cents for each
addition hour or fraction of an hour. For a car parked from 10:45 in
the morning until 3:05 in the afternoon, how much money should be
charged?

	17's	Adults
Acceptable Responses		
$1.35, 1.35, 135¢	47	57
Unacceptable Responses		
$1.10, $1.25, $1.60, etc.	46	38
I Don't Know	5	5
No response	2	0

Although the difficulty of RP01 and RC26 differ substantially for
the two older age groups, the percent of "No response" and "I don't
know" responses were relatively stable. This suggests that neither
the 17-year-olds or the adults were intimidated by either problem.
This willingness to attempt these exercises is good and is no doubt
due in part to the natural context in which they were written. That
is, the problem-solving situations were realistic enough to encourage
bona fide attempts at solutions. However, only approximately half of
the 17-year-olds and adults offered a correct answer or used an appropriate solution process on either exercise. These are disappointing
results, and it does not seem unreasonable to expect higher performance
from both age groups.

These rather low results were mirrored in the results for 17-year-
olds and adults on C20027. This unreleased exercise presented a taxi
problem in which the rate was 60¢ for the first 1/5 mile and 10¢ for
each additional 1/5 miles. The mileage for the trip was given and
respondents were asked to determine the total cost. About 10 percent
of the 17-year-olds and less than 20 percent of the adults determined
the correct cost, with over 20 percent of the 17-year-olds and 25 of
the adults responding "I don't know." This exercise seems to have
face value as a "practical" problem. Few taxi riders, however, ever
make such calculations. It is in fact a rather nonroutine problem.

The findings from RC26 and C20027 illustrate that a sizeable portion of these age groups are not skilled in solving such problems.
It is encouraging that adults had a higher level of performance than
17-year-olds. It seems likely that their actual experience with
problem situations was helpful. Similar opportunities to learn could
be provided within regular mathematics classes by carefully developing
and selecting appropriate consumer problems.

RP04 (Table 7.7) dealt with computing federal income tax. Adults
were given certain conditions (such as amount of adjusted gross income,
and the information that this was to be a joint return with three
exemptions) and were asked to determine the amount of federal tax by
reading from a copy of a tax table. Table 7.7 shows that only 55 percent of the adults determined the correct tax; 13 percent selected
"I don't know" in response to this exercise.

Table 7.7
Exercise RP04 and Results

A married couple with an adjusted gross income of $4,588 claim three
exemptions on their federal income tax return, and elect to file a
joint return. According to the tables below, how much tax must they
pay? (Tax tables provided).

	Adults
Acceptable Response	
$215, 215	55
Unacceptable Response	
$318, 225, 208	31
I Don't Know	13
No response	1

An analysis of error patterns for RP04 reveals the most popular
error (10 percent of incorrect responses) was made by reading from the
tax table constructed for two rather than three exemptions. Other
frequent errors result from computing tax for "Single" and "Head of
Household" rather than the "Married, filing joint return" and determining the tax for some other adjusted gross income.

Correct use of a tax table requires careful reading of directions
and a thorough analysis of the table format; consequently, proper
utilization of the tax chart requires careful scrutiny. That it is
easy to make mistakes is evidenced by the several errors that were
detected and by the relatively low percentage of correct responses to
the exercise.

Are problems requiring interpretation from tax tables inherently
difficult? Several unreleased exercises were examined to determine if
reading other tax tables was an equally difficult task. For example,
exercise P00001 required 17-year-olds and adults to determine the
amount of sales tax for specific amounts of money. A typical tax
schedule was provided. The respondent had to locate the appropriate
cost interval and read the corresponding tax. About 80 percent of
each age group answered this exercise correctly. Exercise P20004

required similar use of a tax table in connection with a catalog order, and was administered only to the adults. Approximately 70 percent of this age group determined the correct tax. High performance levels on these exercises suggest that 17-year-olds and adults can generally read simple tax tables; however, their levels of performance are directly influenced by the complexity of the tables.

The practical significance of the federal income tax problem demands serious attention with mathematics classes. Careful instructional planning designed to incorporate this particular consumer problem into the mathematics curriculum seems not only appropriate but essential.

Exercise RP07 (Table 7.8) presented a typical purchasing problem that required the consumer to determine the best buy on different quantities of rice. Approximately one-fourth of the 13-year-olds and less than 40 percent of the two older age groups answered this problem correctly. In fact, this was the most difficult consumer exercise for each of the age groups. Nearly half of each age group selected the largest package. The widespread belief that "if you buy more, it costs less" may have influenced this choice. On the other hand, determining the lowest price per ounce required computing to the hundredth of a cent to distinguish between the third and fourth choices.

Table 7.8
Exercise RP07 and Results

A housewife will pay the lowest price per ounce for rice if she buys it at the store which offers:

	13's	17's	Adults
● 1 pound, 12 ounces for 85 cents	25	34	39
○ 12 ounces for 40 cents	13	10	4
○ 14 ounces for 45 cents	9	8	5
○ 2 pounds for 99 cents	46	46	47
○ I Don't Know	6	3	4

A similar consumer problem was individually administered to 17-year-olds and adults. In this unreleased exercise (P20003), pictures of different sized cans were presented. The weight in ounces and the price in cents were also stated. When asked to estimate which would cost the least 46 percent of both age groups chose the heaviest as having the lowest per unit cost.

Subsequent questions were asked to determine how this decision could have been made. Questions involving estimation of cost per ounce showed that approximately 60 percent of the 17-year-olds and roughly 65 percent of the adults could estimate (within one cent) the per ounce cost for each can. Nearly 80 percent of these estimates were made by rounding the weight to the nearest whole ounce before dividing.

When asked to calculate the cost, approximately 80 percent of each age group used a correct process, namely dividing the cost by the weight. However, the percent of correct responses decreased as the complexity of the problem increased. More specifically, when fractional rather than whole numbers of ounces were involved the percent of correct responses decreased by approximately half for both age groups.

Results from exercises RP07 and P20003 may suggest that 17-year-olds and adults are naive buyers, with nearly half of them tacitly assuming that larger sizes cost less per unit. The individually administered exercises showed they could make reasonable estimates of per unit costs upon requests. However, these data showed that when faced with a buying decision, nearly half of both age groups opted for the largest size. These results not only provide evidence to support the need for unit pricing of products, they also suggest that exposure to similar problems within the mathematics curriculum is needed to eliminate the idea that larger quantities always cost less.

Exercise P00002 presented a problem that required the completion of a job time card. Results of this individually administered exercise are reported in Table 7.9. Less than one-third of the 17-year-olds and slightly under half of the adults correctly completed the time card and computed the appropriate amount of pay.

Table 7.9
Exercise P00002 and Results

Imagine that you have a part time job. The time card you have shows the hours you worked during one week. You are paid two dollars and seventy cents an hour. Calculate what your pay would be before taxes, Social Security, and any other deduction.

	17's	Adults
Acceptable Responses	31	48
Unacceptable Responses	66	50
I Don't Know	2	2
No response	1	1

The low percent of no response and "I don't know" responses for each group suggest that nearly all of the respondents had a problem-solving strategy, although for the most part such strategies resulted in incorrect solutions. An examination of unacceptable responses showed that over 40 percent of the respondents within each age group used a correct method (i.e., attempted to find the total number of hours worked and then multiplied by the hourly pay) but made mistakes in the process. This finding suggests that the level of performance on such consumer problems could be increased considerably by a simple reduction in basic computational errors.

The final exercise (P20004) was individually administered to the adults, and involved the completion of a catalog order form. The respondent was orally given necessary information for certain items (such as name of merchandise, quantity, size, catalog number and unit price) and asked to record these data in appropriate places on the form. An examination of the completed forms showed that no more than 70 percent of the respondents filled out any single line (#1, 2, 3 on order form in Table 7.10) entirely correctly. This means that vital information necessary for filling an order was often either misrecorded or omitted.

Table 7.10
Exercise P20004 and Results

This is an order blank for a department store. Imagine that you work in this store. The store receives telephone calls from people who want to order things from their catalog. Your job is to write down on this order blank what each person wants to order and to calculate how much each person's entire order will cost.

	Adults
Acceptable Response	45
Unacceptable Response	52
I Don't Know	1
No response	2

Although only 45 percent of the adults determined the correct total cost for the order, 45 percent of the unacceptable responses had a correct total for the entries in the merchandise and sales tax lines.

The results of P20004 showed relatively low performance on the task of completing an order form for adults. A surprisingly high error rate of 30 percent occurred in the process of translating an oral order to a written form. Actually the results related to deter-mining total costs were very encouraging with over 90 percent using a correct process, although less than one-half of the adults completed this order correctly.

Summary

Purchase and cost problems were generally difficult with percentages of correct responses for adults and 17-year-olds slightly below 60 percent and 50 percent respectively. Typical consumer problems requiring the application of two operations were the most difficult.

Skills in reading and interpreting sales tax tables were high for both 17-year-olds and adults. However, only slightly over half of the adults were able to correctly use a federal income tax table with over 10 percent of them responding "I don't know." The practical significance of this problem requires a high level of competency and demands serious attention within mathematics classes.

Consumer problems related to determining the best buy were surprisingly difficult, and may reflect the belief that "if you buy more it costs less." Interesting enough approximately 60 percent of both the 17-year-olds and the adults could (upon request) estimate within one cent the unit cost and more than 80 percent used a correct process (i.e., dividing the cost by the weight) to calculate unit cost. These results suggest that more respondents made their decision to buy without either estimating or calculating. One implication for mathematics programs is that reasonable skills in estimating need to be developed and practiced in consumer contexts.

Implications for Instruction

Continuous gains in performance were made from the 13-year-olds to the adults. The most dramatic gains were made from the 13- to the 17-year-old group, which was expected due to the direct influence of the mathematics curriculum. However, adults performed consistently higher than 17-year-olds on all types of consumer exercises whether percents, graphs, tables or consumer problems were involved. These gains may simply be the result of maturation and experience in solving consumer related problems. Nevertheless, these consistent differences raise questions regarding current mathematics programs. In particular, are contemporary mathematics curricula placing less emphasis on consumer related problems? It is possible that we are preparing students with significant knowledge gaps, such as consumer problem-solving skills, in their mathematical development? This assessment in mathematics clearly does not answer these questions, but it may help bring them into focus.

The overall low performance on consumer mathematics exercises provides a strong rationale for including various types of problems in the mathematics curriculum. Clearly not all exercises should be

consumer related nor should such exercises necessarily be done daily.
A more realistic approach suggest that the overall mathematics curriculum reflect a proper balance of consumer and nonconsumer related problems. Furthermore <u>all</u> students should be provided experience in
solving consumer related problems.

A currently prevailing notion is that consumer problems should be
relegated to junior high, general mathematics or senior business
mathematics while academic topics are explored within the college bound
stream. This plan is totally unacceptable and should be rejected.
These assessment data do not show how students with different mathematics backgrounds fared. Yet the results strongly suggest that low
performance is not limited to low achievers, but is generally widespread
among 17-year-olds and adults. <u>Consequently, serious consideration
should be given to the careful selection and systematic integration of
relevant consumer problems throughout the entire mathematics curriculum.</u>

VIII

Other Selected Topics:
Sets and Logic; Exponents;
and Probability

Sets and Logic

<u>Overview of Results</u>

Although there were few released assessment exercises that
dealt directly with the topics of sets and logic, overall results
were fairly encouraging. The small number of exercises administered (both on sets and logic) makes it difficult to draw any
definite conclusions, but following are some general observations
that may be made.

    I.   About half of the 9-year-olds were able to determine
           the cardinality of a given set.

   II.   About half of the 13-and 17-year-olds could form the
           union of two sets, but the responses of the 9-year-
           olds and adults indicated that most respondents in
           these age groups were unfamiliar with this concept.

  III.   Performance levels for 13-year-olds increased from
           59 to 88 percent correct when a Venn diagram was
           used to illustrate set intersection. A similar
           increase from 20 to 45 percent correct was noted
           for the 9-year-olds. Over 70 percent of the 17-
           year-olds were able to form the intersection of
           two sets.

   IV.   Relating set operations to written sentences was a
           fairly difficult task for both 13-and 17-year-olds,
           with fewer than one-third and one-half of these
           age groups respectively responding correctly.

    V.   A majority of all age groups could successfully apply
           the transitive property in order to make comparisons
           in certain problem situations.

   VI.   About half of the 17-year-olds and adults were able
           to recognize logical equivalents of a verbal state-
           ment.

<u>Sets</u>

Nine exercises, four of which are released, dealt with set
concepts. All but one exercise dealt with simple applications of
definitions of set, element, union and intersection.

In one exercise (RD02) 9-year-olds were asked to find how
many elements there were in the set $\{6,3,2,7\}$. Forty-nine per-

cent correctly identified the cardinality of the set, nine percent added the numbers, and 19 percent either did not respond or
responded "I don't know." The rest gave some other incorrect
response.

Few 9-year-olds or adults demonstrated any familiarity with
set union while about half of the 13-and 17-year-olds correctly
found the union of two sets. (See Table 8.1.) Over 15 percent
of the 13-and 17-year-olds found the intersection of the two sets
rather than the union, while the most common error for 9-year-olds
and adults was to add the elements in the two sets. Results of
two unreleased exercises indicate that the percent of correct
responses is about the same when the word "union" was used instead
of the symbol.

Table 8.1
Exercise RD04 and Results

Given $A = \{2,4,5\}$ and $B = \{1,2,3,6\}$, what is A∪B?

	9's	13's	17's	Adults
Correct, with Braces	3	37	50	3
Correct without Braces	2	6	6	2
Repeated Elements e.g., $\{2,4,5,1,2,3,6\}$	3	2	1	2
6 or 7	2	0	0	0
21 or 23	11	4	3	8
A∩B	1	18	15	3
Other unacceptable	19	15	7	10
No Response or I Don't Know	60	19	19	71

On two unreleased exercises 20 percent of the 9-year-olds, 59
percent of the 13-year-olds, and 71 percent of the 17-year-olds
correctly found the intersection of two sets. Fewer than 5 percent
erroneously found the union instead of the intersection. The 16
percent gain for 13-and 17-year-olds over their performance on
exercise RD04 is interesting in that this is almost exactly the
percent that calculated the intersection when directed to find the
union. Thus, it appears that about 15 percent of both age groups
find the intersection any time they are asked to operate on two
sets. Pictures of set relations facilitated the calculation of
intersection considerably, as 45 percent of the 9-year-olds and
88 percent of the 13-year-olds found the intersection of two sets
pictured in a Venn diagram.

On the other hand, 13-and 17-year-olds encountered significant difficulty in relating set operations to written sentences
(see Table 8.2).

Table 8.2
Exercise RD03 and Results

If R is the set of all red cars and S is the set of all sports cars, which one of the following describes the set of all red sports cars?

	13's	17's
○ The complement of R	14	6
○ The complement of S	6	4
○ The union of R and S	48	42
● The intersection of R and S	29	44
○ I Don't Know	3	4

Responses of 17-year olds were about evenly divided between union and intersection as the set operation that described the conjunction of the properties red and sports car, whereas a significantly greater number of 13-year-olds favored union over intersection. A possible source of these errors is the fact that union is associated with the idea of joining, as is the word "and", i.e., union describes joining sets; "and" is associated with addition. Therefore, for many pupils the set of cars that are red and sports cars is thought of in terms of addition or union.

Logic

In two of the logic exercises respondents were asked to compare the ages of three individuals. In the exercise summarized in Table 8.3, 13-year-olds were specifically asked to calculate an age, whereas in the other exercises, which is unreleased, they were

Table 8.3
Exercise RN02 and Results

John is four years older than Ellen, and Ellen is 11 years younger than Monica. Monica is 12 years old. How old is John?

	13's
Acceptable Response	71
27	3
15	12
Other unacceptable	12
No response or I Don't Know	2

only required to use the transitive property to order the different ages. Fifty-four, 81, 84 and 75 percent of the 9's, 13's,
17's and adults successfully solved this problem, with the most
common error being the statement that there was insufficient information. This was a simple, straightforward application of the
transitive property. The fact that as many as 25 percent of the
adults could not solve this problem shows the basic properties of
relations do not develop naturally, and sufficient attention must
be devoted to them in order for all individuals to be able to apply
them to make specific comparisons.

In another exercise 17-year-olds and adults were asked to determine the logical equivalence of verbal statements (see Table
8.4).

Table 8.4
Exercise RN01 and Results

Which one of the statements below follows logically from the
statement, "All good drivers are alert?"

	17's	Adults
○ All alert persons are good drivers	24	26
○ Some alert persons are not good drivers	10	7
○ A person who is not a good driver is not alert	14	9
● A person who is not alert is not a good driver	51	56
○ I Don't Know	2	1

The statement was of the form: good driver implies alert or
$G \Rightarrow A$. Over half of each age group recognized that this implied
that Not $A \Rightarrow$ Not G. About a quarter chose a response of the form
$A \Rightarrow G$. It would be dangerous to extrapolate to generalizations
about students' ability to deal with the equivalence of specific
logical forms, as the context of the statement undoubtedly affects
performance. However, suffice it to say that about half of the
17-year-olds and young adults have substantial difficulty recognizing the equivalence of statements of the "If..., then...." form.

Summary

The concept of set union was familiar to about half of the 13-
and 17-year-olds, but only a few 9-year-olds and adults demonstrated
familiarity with this topic. Improved performances were noted on
exercises that dealt with set intersection; the use of Venn diagrams

facilitated performance even more.

A majority of each age group could successfully apply the transitive property in a simple problem situation, and about the same proportion of the older respondents could recognize logical equivalents of verbal statements.

Implications for Instruction

These exercises are of interest as statements of respondents' level of knowledge, but the results convey very few implications for instruction. Some argument could be made for regarding the set items as somehow indicative of performance in new mathematics. No one seriously accepts this.

The logic exercises were probably answered as much by the semantic context of the problems as by any knowledge of logic. More information is needed in order to draw any implications.

Exponents

Overview of Results

Twelve exercises, of which five were released, were related to exponents. Nearly all of them involved either numerical computation, or interpretation from square root tables. Following are the major observations that may be drawn from the data:

I.   A majority of 13-and 17-year-olds were able to evaluate an expression that involved a positive integral exponent, but negative, fractional, and zero exponents presented much greater difficulty.

II. Approximately two-thirds of the 13-year-olds were unsuccessful on exercises that required knowledge of square roots for their solution.

III. A majority of 17-year-olds and adults were successful on some exercises that involved square root, but fewer than one-fifth of these age groups were successful on tasks that required use of a table of square roots.

Concepts and Properties of Exponents

Three exercises dealt with the concept of integral exponents. Results from Exercises RF01 and RF02 are reported in Tables 8.5 and 8.6. Table 8.5 shows that half of the 13-year-olds and three-fourths of the 17-year-olds have the basic notion of interpreting exponents and raising a number to a positive integral power. However, with regard to zero exponents, less than 20 and 30 percent of the 13-and 17-year-olds respectively could apply the basic definition of x^o , where $x \neq 0$, as required by Exercise RF02.

Table 8.5
Exercise RF01 and Results

$4^3 =$ _____

	13's	17's
Acceptable Response		
64 or 4 x 4 x 4	50	74
Unacceptable Response		
12, 48, and others	36	20
I Don't Know	12	5
No Response	2	1

Table 8.6
Exercise RF02 and Results

$3^0 =$ _____

	13's	17's
Acceptable Response	17	28
Unacceptable Response		
0	34	35
3	32	25
Other	9	3
I Don't Know	6	7
No Response	2	2

An unreleased exercise, F30021, was similar in format to RF01 and RF02, except the exponent was a negative integer. Only 20 percent of the 17-year-olds responded correctly to this exercise.

Another fundamental property of exponents was examined in two unreleased exercises. Problems of the form a^c x b^c and $a^c + b^c$, where a, b and c were positive integers were proposed. The percent of correct responses for the problem involving a^c x b^c was predictably higher than the percent of correct responses for the form $a^c + b^c$, although the performance level for both of these exercises seems satisfactory for both 13-and 17-year-olds.

RF05 was the only exercise that involved a fractional exponent other than one-half (see Table 8.7). It was difficult for the 17-year-olds, as only 19 percent answered correctly and nearly one- fourth responded "I don't know." Analysis of response patterns shows that translating $27^{1/3}$ to $27 \div 3$ (or equivalent) was done by 29 percent of the 17-year-olds. This error suggests that many respondents misread 1/3 as a factor or else they do not understand fractional exponent notation.

Table 8.7
Exercise RF05 and Results

$27^{1/3}$ = _____

	17's
Acceptable Response	19
Unacceptable Response	
27 ÷ 3 (or equivalent)	29
9 x 27 or 243	2
Other	23
I Don't Know	23
No Response	4

Five of the 12 exercises that involved exponents required knowledge of square roots for their solution. RF03 represents the easiest of these exercises and is shown in Table 8.8. Just over one-third of the 13-year-olds and over half of the older two age groups responded correctly. The large number of "I don't know" responses made by the 13-year-olds and adults was somewhat surprising when viewed in light of the fact that the topic of square root is usually included in most mathematics curricula.

Table 8.8
Exercise RF03 and Results

What is the SQUARE ROOT of 16?

	13's	17's	Adults
Acceptable Response	37	75	60
Unacceptable Response			
8	13	5	4
256	4	2	2
Other	15	7	6
I Don't Know	28	9	27
No Response	4	3	1

Exercises RF04, F30006 and F50003 also dealt with the concept of square root. Two of the exercises involved the calculation of square roots and the third required use of a square root table. In only one of these exercises (RF04) did the percent of correct responses for any age group reach fifty. Exercise F50003 presented a standard table of squares and square roots. In F50003A respondents were asked to find the square root of a number (N). Although the problem required a slight manipulation, the answer could have been read directly from the table. Even so, less than 20 percent of either the 17-year-olds and adults responded correctly. F50003B asked for the square root of 100N. Given this task, the rate of correct responses for both groups dropped to approximately 10 percent.

Summary

The majority of 13-and 17-year-olds appear to be able to operate with positive integral exponents, but negative, zero, and fractional exponents presented problems. The concept of square root apparently was not a well-established one for the 13-year-olds, although the 17-year-olds and adults were successful for the most part on exercises that assessed this concept. However, these older age groups were not able to correctly use tables of square roots.

Implications for Instruction

Interpretation of the results should be tempered by several facts. Few 13-year-olds would have much exposure to exponents as these concepts are usually developed in high school algebra. Furthermore, since not everyone takes algebra, there will be a goodly portion of 17-year-olds with either no knowledge or at best a very limited exposure to exponents. A similar situation exists for adults. Not only may there have been little, if any, exposure to the topic of exponents, the years since formal schooling has ended will no doubt take their toll in simple forgetting.

Despite the above facts, several implications are clear. Mathematics teachers will not be surprised to learn that zero, fractional and negative exponents are troublesome topics and deserve special attention. However, the most significant finding concerns the overall low performance of 17-year-olds and adults on the exercises that required the use of tables of square roots. It may be that many pupils never received instruction in the use of tables of squares and square roots. In fact, the modest increase in the performance of young adults over 17-year-olds in using the square root tables may simply reflect the additional experience the older group has gained in using tables in general.

Mastery of square and square root tables will rarely develop incidentally. Careful and systematic instruction is generally required. Results of these exercises indicate that such instruction is clearly needed. Furthermore, teaching pupils the basic skills

of using tables effectively has long-range payoff that transcends
the calculation of squares and square roots.

Probability

Overview of Results

Nearly all exercises related to probability were presented
through verbal problems. The exercises were simple, developed in
a realistic context, and required little computation. Eleven
exercises (5 released) were used to assess concepts of probability
as well as skills frequently used to solve probabilistic problems.
The three specific skills assessed involved factorials, permutations and combinations. Following are the major observations for
which supporting data will be presented:

I.   The majority of 17-year-olds apparently are
     unfamiliar with the concept of evaluating
     expressions that involve factorials.

II. Fewer than one-third of the 13-year-olds were
     able to respond correctly to an exercise in-
     volving permutations.

III. Combinations and permutations are unfamiliar
     topics to the majority of 17-year-olds and
     adults questioned. Results from exercises
     that were designed to measure respondents'
     notions of the probability of a single event
     were inconclusive.

IV. Nearly one-third of the nine-year-olds and an
     average of two-thirds of the 13-and 17-year-
     olds were able to determine the liklihood of
     the occurrence of an event in one problem
     situation, but percentages of correct response
     on other similar exercises were far lower.

V.   Overall performance of all age groups on
     exercises involving expected values were low.

Factorials, Permutations and Combinations

The average performance on each of these topics for each of
the age groups was low. The factorial problem is unreleased;
however it merely presented an expression using factorial notation and asked the respondents to evaluate the expression. Only
five percent of the 17-year-olds could answer correctly, which
clearly indicates that factorials were recognized and used by
a very small group.

Exercise RJ02 involving permutations is shown in Table 8.9.

Table 8.9
Exercise RJ02 and Results

Three friends agree to change the order in which they go through the lunch line each day. In how many different ways can they arrange themselves?

	13's	17's	Adults
No Response	1	1	0
Acceptable Response (6)	30	47	23
Unacceptable Response (3, 5, 9, others)	64	49	72
I Don't Know	5	2	5

RJ02 was difficult for all age groups, with the 17's scoring the highest and less than half of them answering correctly. About two-thirds of the unacceptable responses were either 3 or 9. The three was probably obtained by simply moving one person into each of the three possible positions. Although the answer is incorrect, it reflects a valid process which was just not carried far enough. One extension of this process could also result in 9. The respondent may have reasoned that if one person can be placed in any of the three positions, then the number of arrangements for three people would be 3 multiplied by 3 or 9. Or 9 might be the result of rotely computing 3 times 3 and thus a mechanical process.

Exercise J11003 presented a problem involving combinations, a special case of permutations. This was a very difficult exercise for both 17-year-olds and adults. An examination of the unacceptable responses showed a wide variety of errors, only two of which occurred over ten percent of the time. One such error, namely multiplying the number in the group itself, occurred over 20 percent of the time for 17-year-olds and young adults. This error is similar to the one discussed in RJ02 and also appears to be a mechanical response. The other frequent error, committed by 15 percent of each age group, seemed to treat the problem as a permutation rather than a combination. Although incorrect, a reasonable strategy was employed.

Probability of Single Event

Probability is a mathematical topic regularly encountered by everyone in one way or another. Several assessment exercises, only two of which were released, presented different problems involving probability. One of the most informative exercises was individually administered to 9, 13-and 17-year olds. Since this exercise was unreleased it can only be generally discussed. Nevertheless, these results will help better interpret performance levels on other probability exercises.

Two dice containing one digit numerals, but not exactly the same numerals as regular dice, were used. Questions similar to those in Table 8.10 were asked:

Table 8.10
Dice Exercise and Results

	9's	13's	17's
Is any one number more likely to occur than any other?	51	79	88
What is the smallest sum?	69	83	87
Identify sample space. (Range of correct responses)	71-90	76-97	79-98
What is the probability of a particular sum (which was specified) occurring?	0	0	3

About one-half of the 9-year-olds and a clear majority of the 13-year-olds and 17-year-olds knew that the faces of a die were equally likely to occur. Respondents within all age groups showed a good basic knowledge of events that could occur. This was shown by the high rate of correctly identifying the smallest sum as well as other points in the sample space. Most errors made in identifying points in the sample space seemed to be the result of previous experience with regular dice. Many respondents had already established a mental set. For example, the sum of one was possible with the dice used in this exercise, yet approximately one-fourth of each age group did not identify this event.

Although basic notions of possible events seemed well-established, the question related to the determination of the probability of a particular event was extremely difficult for all age groups. A correct solution required the determination of the total number of outcomes as well as the number of ways the desired sum could be obtained. Unfortunately there were no data available that provided clues related to this basic knowledge of probability. Perhaps it is sufficient to say that correct solutions were very rare.

Problems involving the direct determination of the probability of an event were difficult (less than 25 percent correct) for all age groups. In order to better appreciate this overall low performance, two typical exercises are shown in Table 8.11.

Table 8.11
Exercises RJ01 and RJ03 and Results

In three tosses of a fair coin, heads turned up twice and tails turned up once. What is the probability that heads will turn up on the fourth toss?

	13's	17's	Adults
No response	3	3	1
Acceptable Response	15	27	36
Unacceptable Response	60	56	47
I Don't Know	22	14	16

There are five black buttons and one red button in a jar. If you pull out one button at random, what is the probability that you will get the red button?

	13's
No response	2
Acceptable	11
Unacceptable	73
I Don't Know	14

Clearly these are simple exercises involving basic notions of probability. These results, together with other similar exercises, show that performance in all age groups is very low. The percentage of "I don't know" responses remained generally stable (10-20 percent) for all age groups. This is a sizeable portion, and worthy of particular concern for the two older groups; however, it also suggests that most respondents had a strategy to solve the problem. The nature of the strategies employed was not clear, but intuition seemed to be one particular factor that was likely favored.

Expected Values

Several exercises involved expectations and expected values. For example, if you drop six coins, how many heads would you expect? Forty-one, 55 and 63 percent of the 9, 13-and 17-year-olds respectively, were correct. Another type of expectation problem is shown in Table 8.12 by Exercise RJ04.

Table 8.12
Exercise RJ04 and Results

For four games you have the following chance of gaining points:

    Game A: 10 percent chance of gaining 20 points
    Game B: 20 percent chance of gaining 15 points
    Game C: 40 percent chance of gaining 10 points
    Game D: 50 percent chance of gaining 5 points

In the long run, you would be most likely to gain the GREATEST number
of points in

		17's	Adults
○	Game A	17	16
○	Game B	6	9
●	Game C	31	39
○	Game D	42	30
○	I Don't Know	4	6

Less than 40 percent of either the 17-year-olds or the adults responded
correctly, which is approximately the performance levels for all
exercises related to expected value. RJ04 required the use of percents
in computing the expected value. It was impossible to identify specific
computational errors, however it is likely that some unacceptable
responses could be attributed to lack of skills in working with percent.

Summary

     The overall results of these probability and probability related
exercises were disappointing. High performance was observed only on
questions related to describing a sample space and identifying possible
outcomes. However, the basic concept of the probability of an event
as the ratio of the number of successful outcomes to the total number
of outcomes was clearly lacking. Low performance was also found in
permutations and combinations, but none of the probability problems
required the use of either notion. The probability exercise generally
presented problems which in one way or another are frequently
encountered and the computation required for a solution was minimal.

Implications for Instruction

     Probability does not occupy a major, or even a minor portion of
most mathematics programs. It is usually omitted or treated on rare
occasions -- such as the day before a vacation starts. Consequently
these poor results do not reflect directly on the mathematics programs
and may come as no surprise. Yet it is difficult to identify a topic

in the mathematics curriculum which holds the potential for greater use, excluding computational skills, than basic notions of probability.

These results show there were no consistent differences between 17-year-olds and adults. Furthermore the differences where they did exist were not substantial. It is clear that maturation alone will not develop competence in this area. Consequently, if fundamental notions of probability are to be developed, then these concepts must be an integral part of school mathematics programs, and not considered simply a luxury topic. Probablistic ideas must be developed carefully and presented systematically over a period of time. The area of probability holds a high interest for students. The variety of probabilistic activities, both concrete and abstract, which can be used within a classroom make it ideally suited for a wide range of students.

Epilogue

What have we accomplished in presenting this monograph? Actually someone other than the authors will have to answer the question in the long run. But, there are a few things we would like to mention.

First, the early draft of this monograph helped us develop a framework to prepare the various journal reports that have preceded this publication. We accomplished our task of preparing interpretive reports working directly from the NAEP exercises and the NAEP data printouts. The NAEP technical reports did not precede our work. Thus this monograph served us first as an organization of the entire set of mathematics data from which we developed the more targeted articles in the journals.

Second, by its publication we are making available a reference for the whole mathematics assessment. Other reports and articles have selected age groups or exercises to present more specific discussion. Some exercises and data have not been reported previously by the NCTM project.

Third, mathematics educators are still learning how to deal with assessment data. One of the lessons to be learned is that we really have a situation quite unlike achievement testing. There are no test scores to report, yet people unfamiliar with assessment processes conceive of a "national assessment mathematics test." As we struggled to organize and report this large mass of data, our years of exposure to test scores and test score interpretation surely got in our way. Rather, we had a large set of exercises and percentages associated with various responses. Presenting and interpreting this information is much more complex, but we believe potentially more useful, than presenting and discussing test score data. It is much more like describing student performance than measuring it. We believe this monograph will help mathematics educators better understand the assessment process and its data and to move away from the test score syndrome.

Fourth, the process of preparing this monograph allowed us input into the planning for the second mathematics assessment. Some of our suggestions were incorporated:

--broader input from mathematics educators and teachers in the specifying of objectives and in exercise development

--organizing the assessment, from planning through reporting, around a set of questions based on priorities in mathematics education

--better coverage of key topics with clusters of exercises.

The preparation of the second mathematics assessment did indeed involve a cross-section of the mathematics education community--mathematicians, teacher educators, teachers, and concerned citizens. Exercise development was coordinated by the National Assessment staff, rather than an independent contractor and this in particular enabled them to involve lots of mathematics educators in the planning and development. The questions focussed on performance with computational skills, with understanding mathematics, and with using mathematics. The better coverage was accomplished by use of a larger number of exercises and selection of key topics on the basis of the questions.

Finally, we believe the monograph provides a data base from which to ask questions about mathematics achievement. Such activity forces recognition that some fundamental issues precede interpretation of data. What is satisfactory performance in computational skills? What computational skills are expected of 9-year-olds? of 13-year-olds? of 17-year-olds? of all citizens? What uses of mathematics can be expected of all citizens? These and many other questions must be faced. National assessment and its advisors face them in planning, in exercise development, and in reporting. But, the reader also must deal with such questions as he interprets the information presented and tries to answer, "What do these data say for me, for my school, or my students?"

References

National Assessment of Educational Progress. <u>Math fundamentals: Selected</u> <u>results from the first national assessment of mathematics</u>. 04-MA-01, January 1975.

National Assessment of Educational Progress. <u>Consumer math: Selected</u> <u>results from the first national assessment of mathematics</u>. 04-MA-02, June 1975.

National Assessment of Educational Progress. <u>Mathematics technical report</u>: <u>Exercise volume</u>. 04-MA-20, February 1977.

National Assessment of Educational Progress. <u>Mathematics technical report</u>: <u>Summary volume</u>. 04-MA-21, September 1976.

National Assessment of Educational Progress. <u>The first national assessment</u> <u>of mathematics: An overview</u>. 04-MA-00, October 1975.

Reys, R. E. Consumer math: Just how knowledgeable are U.S. young adults? <u>Phi Delta Kappan</u>. 1976, November, 258-260.

Wilson, J. W., Carpenter, T. P., Reys, R. E., & Coburn, T. G. Notes from national assessment: Addition and multiplication with fractions. <u>The Arithmetic Teacher</u>, 1976, February, 137-142.

Wilson, J. W., Carpenter, T. P., Reys, R. E., & Coburn, T. G. Notes from national assessment: Basic concepts of area and volume. <u>The Arithmetic Teacher</u>, 1975, October, 501-507.

Wilson, J. W., Carpenter, T. P., Reys, R. E., & Coburn, T. G. Notes from national assessment: Estimation. <u>The Arithmetic Teacher</u>, 1976, April, 296-302.

Wilson, J. W., Carpenter, T. P., Reys, R. E., & Coburn, T. G. Notes from national assessment: Perimeter and area. <u>The Arithmetic Teacher</u>, 1975, November, 586-590.

Wilson, J. W., Carpenter, T. P., Reys, R. E., & Coburn, T. G. Notes from national assessment: Processes used on computational exercises. <u>The Arithmetic Teacher</u>, 1976, March, 217-222.

Wilson, J. W., Carpenter, T. P., Reys, R. E., & Coburn, T. G. Notes from national assessment: Recognizing and naming solids. <u>The Arithmetic Teacher</u>, 1976, January, 62-66.

Wilson, J. W., Carpenter, T. P., Reys, R. E., & Coburn, T. G. Notes from national assessment: Word problems. <u>The Arithmetic Teacher</u>, 1976, May, 389-393.

Wilson, J. W., Carpenter, T. P., Reys, R. E., & Coburn, T. G. Research implications of the NAEP year 04 mathematics assessment. Journal for Research in Mathematics Education, 1976, 7, 327-336.

Wilson, J. W., Carpenter, T. P., Reys, R. E., & Coburn, T. G. Results and implications of the NAEP mathematics assessment: Elementary school. The Arithmetic Teacher, 1975, October, 438-450.

Wilson, J. W., Carpenter, T. P., Reys, R.E., & Coburn, T. G. Research and implications of the NAEP mathematics assessment: Secondary school. The Arithmetic Teacher, 1975, October, 453-470.

Wilson, J. W., Carpenter, T. P., Reys, R. E., & Coburn, T. G. Subtraction: What do students know? The Arithmetic Teacher, 1975, December, 653-657.

Wilson, J. W., & Martin, W. L. The status of national assessment in mathematics. The Arithmetic Teacher, 1974, 21, 49-53.